High Above the Alps

A Bird's Eye View of Geology

Kurt Stüwe and Ruedi Homberger

Weishaupt Publishing

VOGES
BLACK FOREST
Kempte
Lake Constance
Allgäu A
Zurich
Lechtal A
JURA
Neuchâtel
Bern
Napf
Glarus Alps
Chur
Davos
Rhine
Arosa
JURA
Fribourg Alps
Eiger
Bernese Oberland
Lukmanier Pass
Lepontine
Ticino
Piz Bernin
Bergell
D. Morcles
JURA
High Savoy
Rhone
Geneva
Simplon Pass
Valais Alps
Bergamo Alps
Mt. Blanc
Mte. Rosa
Aosta Valley
Lago Maggiore
Lake Como
Bauges Massif
Gran Paradiso
Milan
Grenoble
Vercors
La Meije
Ecrin M.
Briancon
Turin
MASSIF CENTRAL
Dauphiné Alps
Monviso
APENNINES
Gap
Cuneo
Genoa
Maritime Alps
Argentera Massif
Durance
Avignon
Verdon
Nice

Linz
Krems
Wachau
Melk
Vienna
Bratislava
Munich
Chiemsee
Salzburg
Salzkammergut
Schneeberg
Lake Neusiedl
Berchtesgadener Alps
Wilder Kaiser
Watzmann
Dachstein
Liezen
Hochschwab
Karwendel
Leoben
Budapest
Innsbruck
Zell am See
Lower Tauern
Großglockner
Hohe Tauern
Graz
Brenner
Saualpe
Lienz
Koralpe
Balaton
Meran
Dolomites
Carnic Alps
Klagenfurt
Maribor
Bolzano
Karavanke
Pohorje
Julian Alps
Steiner Alps
Tagliamento
Adige
Belluno
Udine
Piave
Zagreb
Triest
Padua
Venice
Krk
DINARIDES
ADRIA
Bologna
Zadar
APENNINES

High Above

The Glarus Thrust is the most famous geological structure of the Alps. As part of the "Tectonic Arena Sardona" it became part of the world nature heritage of UNESCO in 2008. The photo shows the thrust in the region of Piz Sardona near the border between the three Swiss cantons Glarus, St. Gallen and Grisons. The thrust plane forms a flat lying knife-sharp boundary between rock types of completely different origin and age. Above the structure the rocks are undeformed Permian volcanics ("Verrucano"), that are about 250 million years in age. Below the structure the rocks are much younger: sediments that were deposited only 50 million years ago into the Penninic ocean ("Flysch"). The thrusting itself occurred at about 30 Ma. Erosion exposed the structure only much later: the valleys flanking the entire range between Vorab and Ringelspitz were incised only in the last 1–2 million years. In the rugged rock needles of the Tschingelhörner the ridge has become so narrow that a gigantic hole has broken through the range: the Martinsloch.

the Alps

A Bird's Eye View of Geology

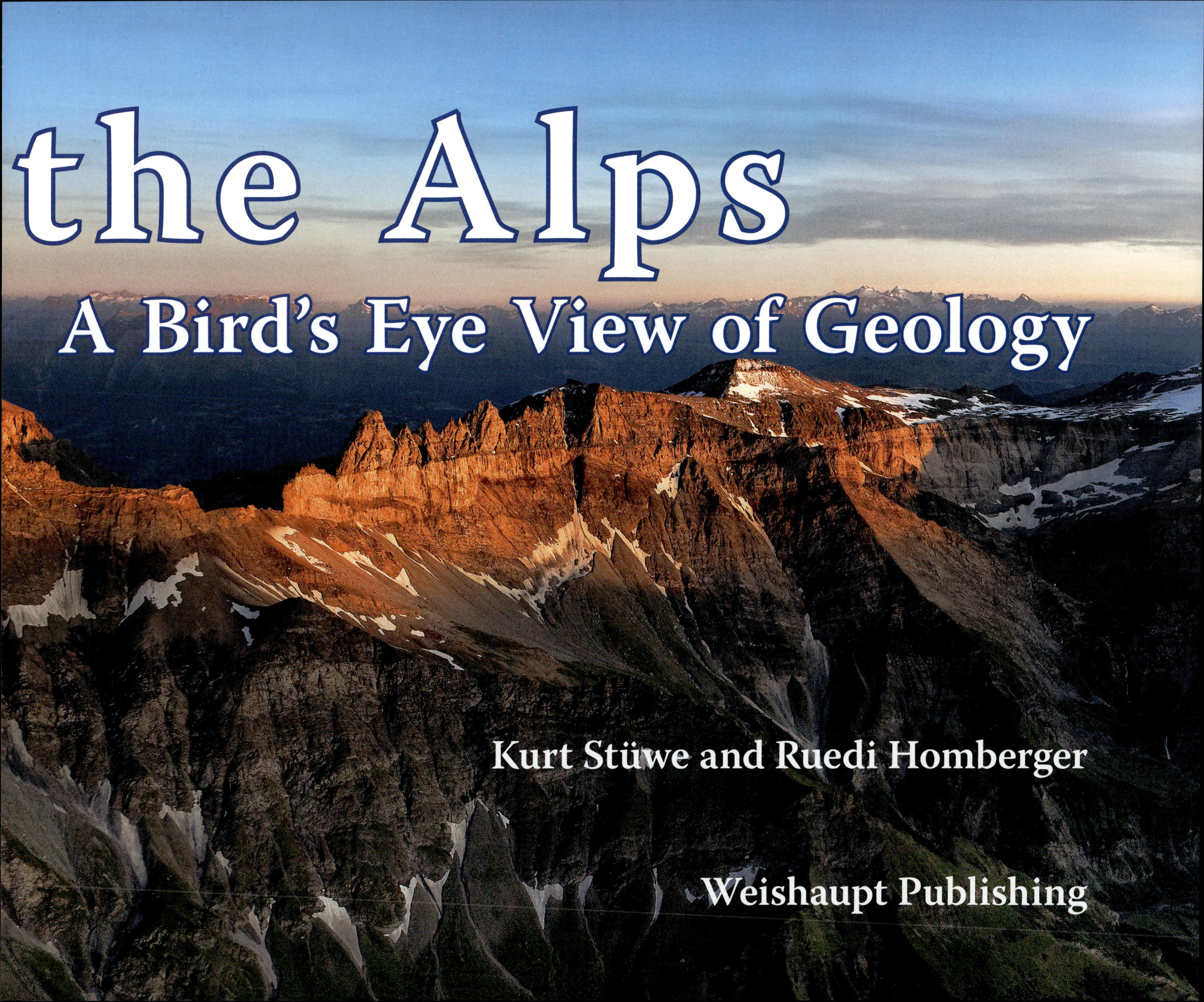

Kurt Stüwe and Ruedi Homberger

Weishaupt Publishing

Mt. Blanc (4 807 m) is the highest peak of the Alps and its impressive grandeur can keep up with the high peaks of any mountain range in the world. Geologically the Mt. Blanc Massif consists of coarse-grained granites and gneisses that form the basement of the Alps and once were part of the Mesozoic European middle crust. In the Alps, these rocks are seen in the so called "External Massifs" which form large dome-shaped structures surrounded by sedimentary rocks that once were deposited onto the Mesozoic European plate. In many places of the Swiss Alps these sediments were later scraped off their base and thrown into surreal folds: these are the Helvetic nappes of Switzerland. Here at the Tête à l'Ane, north of Mt. Blanc, the limestones and marls that constitute these Mesozoic sediments lie flat. Only some million years ago such sediments still covered the top of Mt. Blanc (as indicated by the yellow dome in the sketch at left).

Contents

Left: The Illgraben on the south side of the Rhône Valley near the settlement of Leuk is one of the most actively eroding rock faces of the Alps. It illustrates the dramatic influence of geological processes on the shaping of the Alps. Continuous landslides destroy the massive dolomite and quartzite layers and rapidly consume the thin veneer of soil and forest from the Illhorn mountain.

Acknowledgements

This book is largely a translation of the German language book "GEOLOGIE DER ALPEN AUS DER LUFT", published by Weishaupt Publishing in early 2011. The German version was out of print more than once within the first year of publication, motivating us now to venture into the English market.

For the original German language version, we received enormous help from many colleagues across Europe, providing us with hints for good photo locations, diagrams to accompany the text, coffees at various airfields around the Alps and, in fact, teaching us a lot about the geology of the range – of which we knew embarrassingly little when the project started in 2008. Thank you all for your encouragement and help (for a full listing of friends and colleagues who helped us, see the original German edition).

With its tectonic interpretation of the Alps, the book leans heavily on modern views as taught by the schools of Niko Froitzheim, Stefan Schmid and Ralf Schuster (see overview papers Froitzheim et al., 2008; Schmid et al., 2004 or Schuster and Stüwe, 2011). Thank you for taking the time to discuss the larger geological picture with us!

For the English language edition, we thank Jessica Hedrick for proofreading the text and Jürg Meyer, Neil Mancktelow, Christoph Spötl, Bernhard Grasemann and Jürgen von Raumer for bringing many geological mistakes in the earlier editions to our attention. Karin Ehlers is thanked for finding many little editorial hiccups and, in fact, for structuring our original thoughts on chapter division and layout of the entire book. Urs Homberger spent many hours into archiving and processing the photos reproduced here. Stefan Hergarten and Georg Stegmüller are thanked for help with the project web pages (**www.alpengeologie.org**).

Since the publication of the German language version we have continued to expand our photo database with many more hours of flying and photography. Martin Kennedy and Karin Ehlers are thanked for providing logistic support during these flights. Some of these photos are included on additional pages in this English language edition. SPmining is thanked for financial support (for full credit see p. 284).

A lonely trail on the way to the Muttekopf hut near Imst in Tyrol crosses a spectacular fold made of Cretaceous marine sediments (Gosau sediments, 80 million years old). The photo symbolises the entrance to the world of geology.

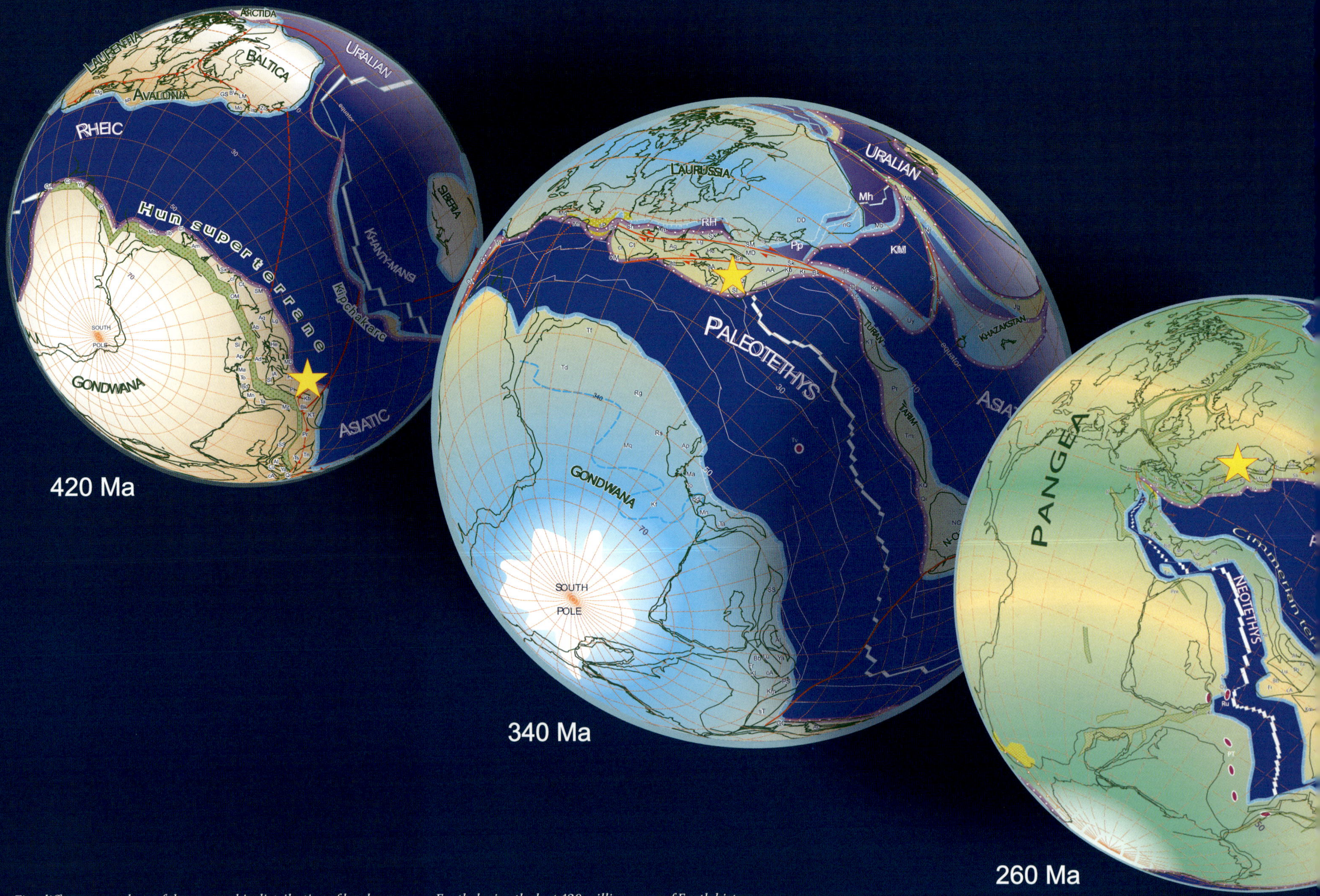

Five different snapshots of the geographic distribution of land masses on Earth during the last 420 million years of Earth history. The yellow star marks the position of today's central Europe. In the Alps, the western part of the Palaeotethys at 260 Ma is called the: "Meliata-Hallstatt Ocean". The domain of the Penninic oceans may be seen on the globe for 121 Ma to the north of the yellow star.

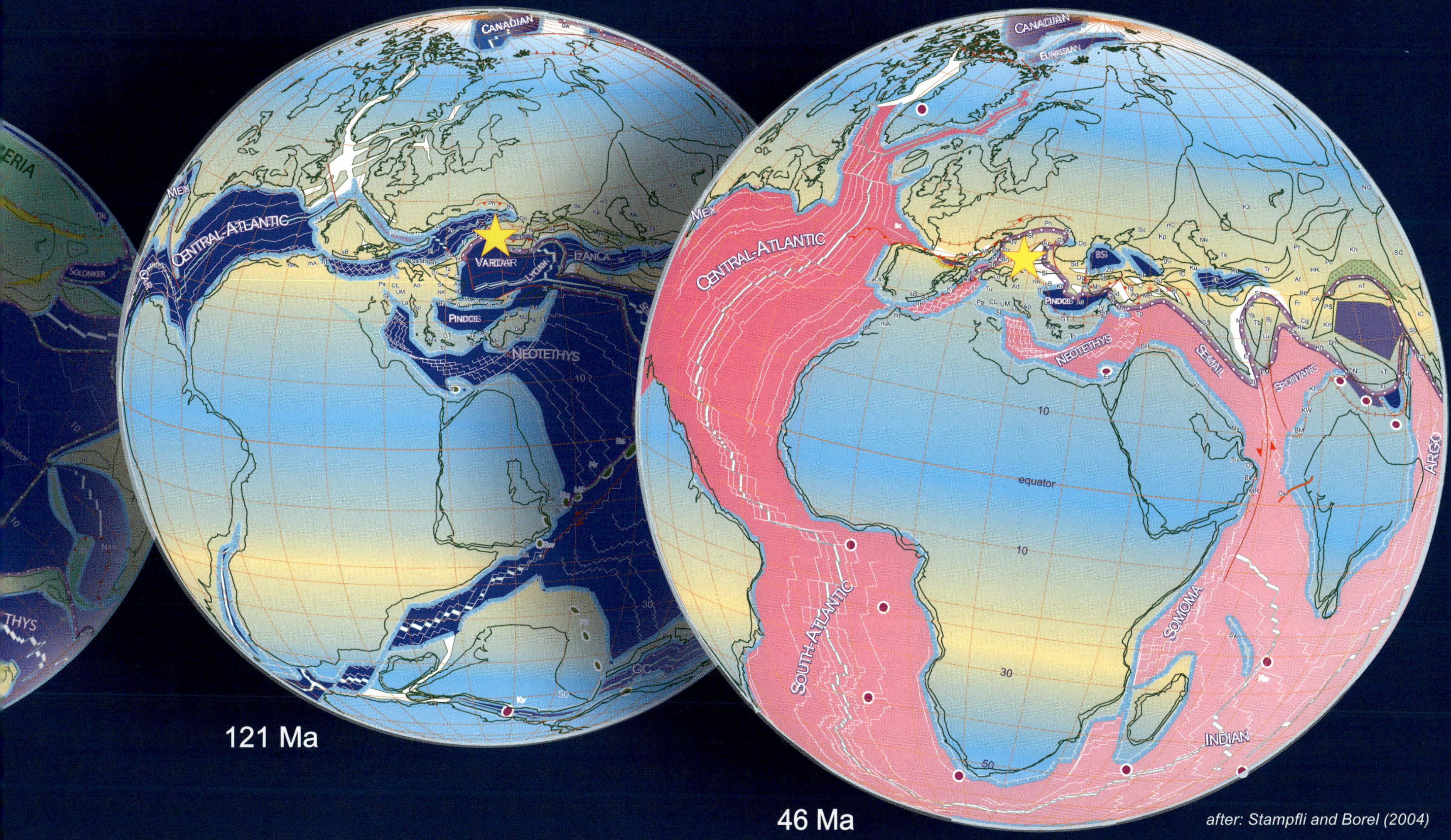

after: Stampfli and Borel (2004)

1. Introduction

The Alps are an astounding mountain range. For about 400 million years their rocks were broken, kneaded and folded at large depths and high temperatures, destroyed by erosion and their sediments re-deposited and re-consumed into new mountain forming processes. At least three times in the known history of the range, entire mountain building cycles (*orogenies*) occurred. Only the last of these three cycles is responsible for the present day geometry of the range. This last orogeny – the *Alpine mountain building cycle* – has formed the rocks and structures of the Alps for about 100 million years. However, the Alps as a topographically elevated mountain range are much younger than this. They uplifted only in the last few million years.

In each of the three mountain building cycles that affected Europe in the last 400 million years, entire oceans were pushed deep into the Earth's mantle, mountain ranges formed and small continents completely squashed. The geological build up of the Alps is the complex result of this enduring process. Geologists like to consider the Alps to be: *"the most complicated mountain range in the world"*. Whether this is true or only our perception is difficult to judge, but the amazing wealth of geological investigations of the Alps has certainly both: provided answers and posed more questions. In any case, the beauty of Alpine geology has caused much awe in the layman and provided a near infinite reservoir of exciting scientific problems for the geologist. The present book is a homage to the fascinating record of the evolution that is preserved in the rocks of the mountain range.

Chapter 1 summarises some of the basic concepts. The *geological clock* is explained to illustrate how apparently slow geological processes may cause the displacement of entire mountains and oceans.

The age of the Alps is then explained in comparison to the age of the Earth as a whole. In this context it needs to be said that the word "Alps" has a different meaning to the geologist than to the climber. For the climber, the Alps refers to the mountains. For the geologist, the Alps include the entire mountain building process with all its structures and rock types of the entire Earth's crust. A short summary of the concept of plate tectonics helps the reader understand how ocean floors can sink into the mantle and continents can collide. The remainder of Chapter 1 introduces the forces that build up and destroy mountain belts: the *endogenic forces* that deform the crust internally and uplift its surface, and the *exogenic forces* that erode and destroy the mountains from the outside and ultimately lower the surface.

Chapter 2 introduces the different building blocks that built up the Alps over hundreds of millions of years from a number of jigsaw puzzle pieces. In total geologists recognise three different continents and two different ocean floors that today lie on top of each other like a giant sandwich. This *Alpine sandwich* includes the crustal fragments of the *Adriatic plate,* the *European plate* and the rocks of the ocean floors that once lay in between. The layered sandwich is discussed in Chapter 2 from bottom to top.

Chapter 3 places the jigsaw puzzle pieces together in terms of a geological evolution. Geologists call this game *palinspastic-* or *palaeogeographic reconstruction*: The art of reconstructing the geographic distribution of land masses and oceans in the geological past. This discussion commences at a time that is about 300 million years ago (*Variscan* time) and several pages are dedicated to each geological period. The influence of the glaciation periods and mankind is also discussed in this chapter.

The landscape around the Muttsee lakes in the Tödi region of the Glarus Alps in Switzerland looks like it may not be on Earth. It illustrates why geology is such a faszinating subject. The vertical limestone pinnacles once were deposited in the Mesozoic as sediments on the European continent. Since then they were thrust up, folded and are now being shaped by wind, water and glaciers.

The Muttsee lakes are also an example of the enormous influence of mankind on the shaping of the Alps. The lakes will soon be connected with a power plant some 2000 m below in the Limmerntal and serve as a reservoir for one of the largest hydroelectric schemes in Europe. Will mankind destroy this unique landscape more rapidly than all geological forces may?

Chapter 4 introduces selected geological highlights of Alpine geology. About 100 of the most spectacular geological formations are illustrated and explained in a geographic order from east to west. For this kaleidoscope of geological highlights, the Alps are divided into four different provinces that are geologically and geographically different: (i) the *Eastern Alps* from the Pannonian plains to the Rhine Valley, (ii) the *Central Alps* from the Rhine Valley to Mt. Blanc, (iii) the *Western Alps* all the way to the Côte d'Azur and (iv) the *Southern Alps* south of the Periadriatic Line, the line that forms today the plate boundary between the European and the Adriatic plates.

A glossary provides explanations of some geological terms used within the text. An index, a literature list and a brief account of the making of this project round off the book at the end.

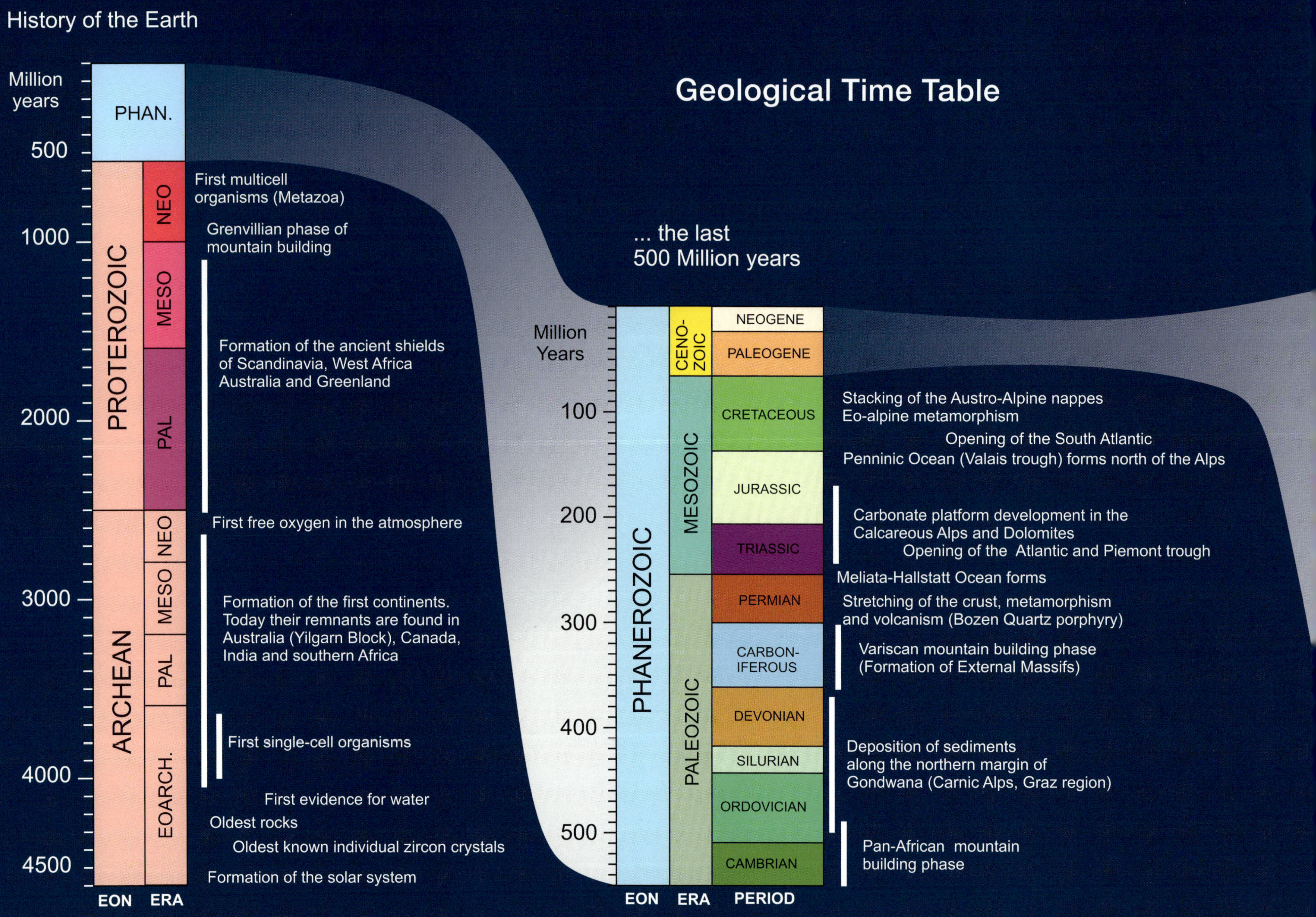
History of the Earth
Geological Time Table
Million years
500
1000
2000
3000
4000
4500
PHAN.
PROTEROZOIC
ARCHEAN
NEO
MESO
PAL
NEO
MESO
PAL
EOARCH.
EON
ERA
First multicell organisms (Metazoa)
Grenvillian phase of mountain building
Formation of the ancient shields of Scandinavia, West Africa Australia and Greenland
First free oxygen in the atmosphere
Formation of the first continents. Today their remnants are found in Australia (Yilgarn Block), Canada, India and southern Africa
First single-cell organisms
First evidence for water
Oldest rocks
Oldest known individual zircon crystals
Formation of the solar system
... the last 500 Million years
Million Years
100
200
300
400
500
PHANEROZOIC
CENO-ZOIC
MESOZOIC
PALEOZOIC
NEOGENE
PALEOGENE
CRETACEOUS
JURASSIC
TRIASSIC
PERMIAN
CARBON-IFEROUS
DEVONIAN
SILURIAN
ORDOVICIAN
CAMBRIAN
EON
ERA
PERIOD
Stacking of the Austro-Alpine nappes
Eo-alpine metamorphism
Opening of the South Atlantic
Penninic Ocean (Valais trough) forms north of the Alps
Carbonate platform development in the Calcareous Alps and Dolomites
Opening of the Atlantic and Piemont trough
Meliata-Hallstatt Ocean forms
Stretching of the crust, metamorphism and volcanism (Bozen Quartz porphyry)
Variscan mountain building phase (Formation of External Massifs)
Deposition of sediments along the northern margin of Gondwana (Carnic Alps, Graz region)
Pan-African mountain building phase

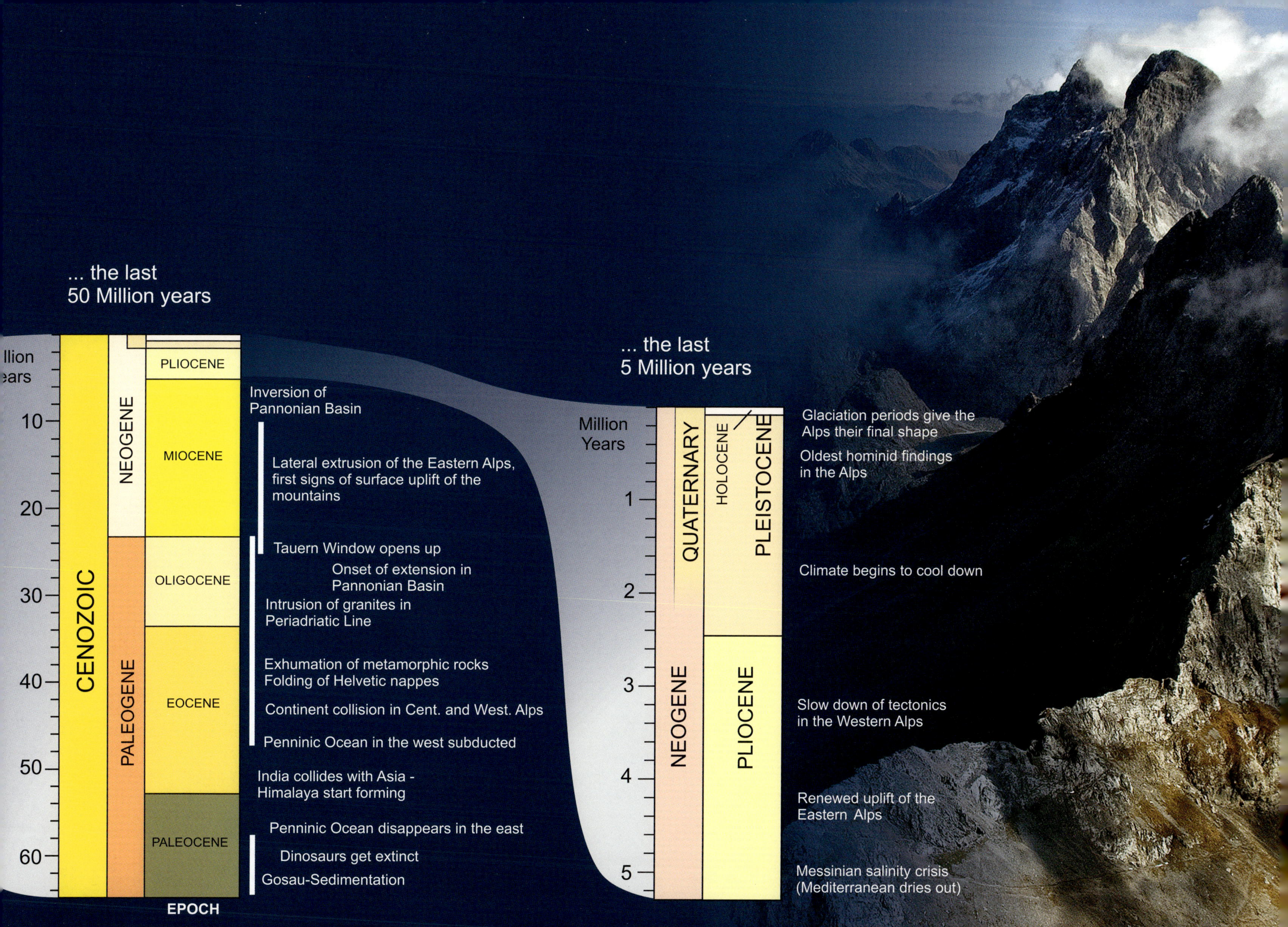

Photo at right: The Hohe Warte (2 780 m) and the Wolayer Lake in the Carnic Alps. This mountain contains the oldest rocks of the Alps (see p. 94) and is famous for a unique juxtaposition of coral reef- and deep water limestones.

1.1. Space and Time Scales in Geology

Life that has become rock! The ammonites here are almost 200 hundreds million years old and look like they still lived yesterday. Rocks like this have always inspired geologists to think about geological time scales.

In order to get a good feel for the magnitude of geological processes it is important to be very generous with the handling of space and time.

Geological Time Scale: As human beings we are used to measuring time with our wrist watches and a calendar. As such, the shortest measured time spans measure seconds, about a year is a manageable time span, but anything beyond about 1000 years is incomprehensibly long. In tectonics, time scales are often measured using the lifetime of a mountain range for scale. Geologists call this a *mountain building cycle* or: *orogenic cycle*. The life time of such a cycle is about one million times as long as the life time of a human being: it takes roughly 100 million years from the beginning of uplift of a mountain range until it is eroded and ultimately planated again. Thus, on a plate tectonic scale, one million years is a short geological time span, and only time spans of billions of years are considered to be long, even on a plate tectonic scale.

The enormous number of geological processes that can be packed into the lifetime of a mountain belt may be understood if the lifetime of a mountain belt is compared with the speed of geological processes. Continental plates drift on the surface of the soft asthenosphere at a rate of some millimetres to few centimetres per year. This is quite well known from a series of different observations. Today it is even possible to measure this directly using GPS methods. Extrapolated, this corresponds to several tens of kilometres in a few million years. It is therefore easily possible that plates drift many kilometres towards each other, overthrust each other and bring rocks from the surface deep into the lower crust and back up again – all easily within a single orogenic cycle.

Erosion by the forces of wind and weather operates at a comparable rate: the sediment load suspended in the large Alpine rivers carry many millions of tons of rock out of the Alps each year. On average, several tenths of millimetres are eroded off the surface of the mountain range each year. In several million years, many kilometre-thick layers are taken off the range. In response to this unloading from above, the lithosphere pops up from the soft Earth's mantle like a cork from water. Therefore a mountain range does not necessarily loose elevation when it erodes.

In short: the time available in a single orogenic cycle is well sufficient to accommodate huge distances of displacement, folding, burial and exhumation of rocks. A large degree of imagination is therefore required in order to understand the appearance of the Alpine geography in geological times!

Geological Spatial Scale: When talking about the geological evolution of the Alps it is usually the three dimensional nature of plate tectonic motions that is most difficult to imagine. Interestingly, the vertical motions of rocks are at least as important to the formation of the Alps as the horizontal motion. Consider the rock types we find in the Alps: the majority of rocks that form the surface of the Alps are crystalline rocks. The mineralogy and structure of these rocks tells us that the minerals that they are made out of grew at pressures and temperatures corresponding to many tens of kilometres of depth in the crust. This is because burial and exhumation, extension, erosion, deformation and sedimentation all move rocks in the Alps roughly at the same rate upwards or downwards as they move them around horizontally.

The Earth's surface, with all its mountains and valleys can be considered as a more or less random horizontal cross section through the Earth's crust. Below this level geologists can observe the rocks. Above this level the rocks were already eroded in the past.

The elevation of the surface above sea level, with all its peaks and valleys, depends on the interaction of tectonically driven uplift and erosion. Curiously enough, the simple question of whether the Alps are rising today or not is one of the most difficult problems in geology.

In summary, the evolution of the Alps may be compared with a *marble cake*. Mix chocolate and vanilla dough (as symbols for the different building blocks of the Alps) and pull a spoon through the mixture until it has brown and white swirls. Then bake it and cut it horizontally through its middle. Voilà: here are the Alps. The little experiment illustrates how difficult it is to interpret where the two different pots of dough were, before the cake was made.

How old is the cake? A good chef will not answer this question. He may only ask back: "What do you mean? ... 'cos the decoration was only done an hour ago: it was then that I cut the cake, decorated it with cream and put the candles on. However, the base was made yesterday and the bread crumbs used for the dough were ground two days ago from old bread rolls made last week. The apples I used for the fruit topping were picked last fall and the glazed almonds on top were brought to me by a friend – I have no idea where they are from ..."

The story of the age of the Alps is quite similar: it is just about impossible to give a simple answer to the question *"How old are the Alps?"*. The appearance of the range today, with its rugged peaks and broad valleys, was only created in the last of the glaciation periods (within the last 100 000 years). The Alps as a mountain range are a few million years old, but already 30 million years ago the collision between the Adriatic and European plates started that formed many of the structures and rocks in the Alps. Just before this collision of these two continental plates the final remnants of the Penninic Ocean (the rocks of which build up much of the Central Alps) were pushed underneath the Adriatic plate and so on ... The further we go back in time, the more events will be discovered. Of course, at the same time we know less and less about them.

Geologists often are asked in the Alps: "Was an ocean once here?" Of course: limestone, marine fossils or deep sea sediments indicate these rocks formed under water. However, just *where* was this water if these rocks were moved for hundreds of kilometres, were thrust over the top of each other, buried, metamorphically changed in mineralogy and chemistry and then squeezed back to the surface?

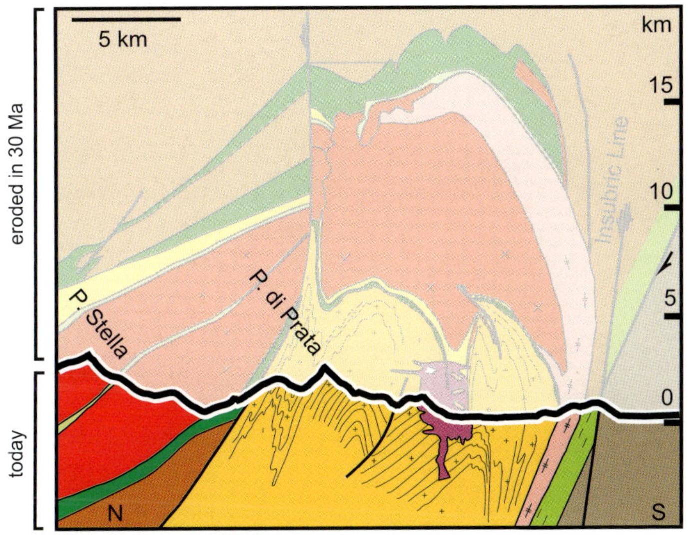

after: Rosenberg (2004)

Geological north-south cross section through the Bergell Massif. The light shaded part of the cross section has long been removed by erosion. The thick black line shows the present day topography. The cross section illustrates that the surface of the Alps is merely a random cross section through the complicated three dimensional geological build up of the Alps. Red = granodiorite; pink = tonalite; green = ophiolite; yellow = Penninic basement; brown = South Alpine; ochre = Austro-Alpine.

1.2. A Crash Course in Plate Tectonics

Different tectonic plate margin processes placed into a schematic cycle that is called the Wilson cycle. From top to bottom these are:

1. Stretching of a continent with formation of a sedimentary basin on its surface (a typical example of this stage is the Rhine rift).

2. Tearing of the continent (rifting) (e.g. Red Sea).

3. Formation of a new ocean (e.g. Atlantic).

4. Subduction of this ocean underneath a continent (e.g. Java, South America).

5. Subduction of a mid-oceanic ridge (e.g. Alaska).

6. Collision of two continents and formation of a new mountain range (e.g. Alps).

In all illustrations the colour code is: orange = continental crust; red = oceanic crust; green = mantle part of the lithosphere; blue = water. The boundary between crust and mantle is called Moho (short for: Mohorovičić-discontinuity). However, the lower boundary of the plates is not the Moho, but the transition between hard rock and soft, convecting asthenosphere (yellow). This is only at a depth where rocks are about 1200°C hot, some 100 km deep in the Earth's mantle.

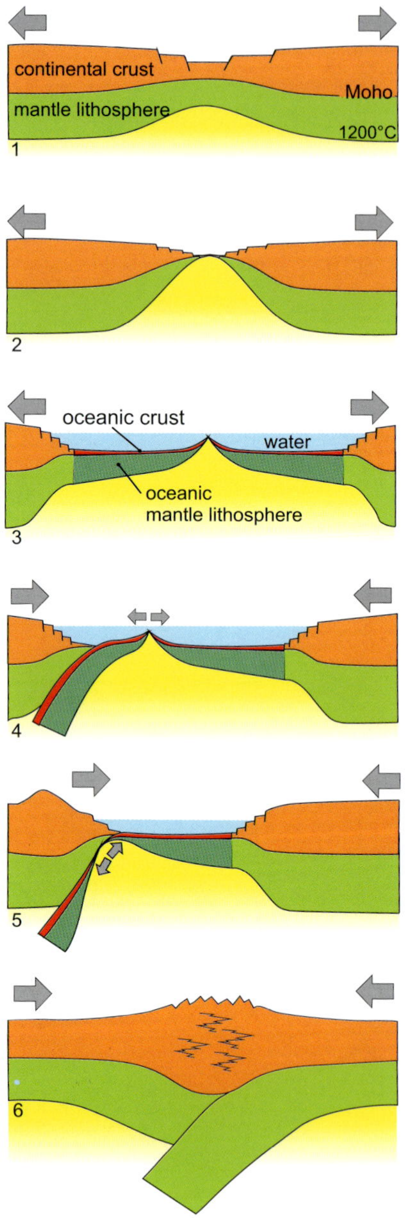

The outer shell of Earth is made of rock and drifts around on this planet as plate-like bodies floating on top of the soft Earth's mantle. This outer shell is called the *lithosphere* (from lithos = rock and sphere = shell). A lithospheric tectonic plate is made up of two parts: the upper part is called the *crust* (in the diagrams at left: brown and red). The lower part is called the *mantle part of the lithosphere* (at left: green), because it is part of the mantle according to its chemical composition, but part of the rocky lithospheric plate according to its physical properties. On Earth there are two types of lithospheric plates, which differ mostly in the build up of their upper part, the crust. These two types are called *continental lithosphere* and *oceanic lithosphere* because their distribution on Earth fortuitously corresponds roughly to the regions covered by land and by sea water, respectively. The crust of the continental lithosphere is normally on the order of 30 – 40 km thick and is made up of partial melts that segregated from the mantle soon after the formation of Earth some four billion years ago. Thus, the chemical composition of continental crust is largely that of granite and the total area of the continents has been about the same for billions of years. In contrast, the oceanic crust consists of basaltic magmas that are constantly being produced from the mantle along the mid-oceanic ridges and is only 5 – 7 km thick. Several square kilometres of oceanic crust are being produced each year even now. This oceanic crust is re-consumed into the mantle every 100 million years or so.

The temperature at the base of the crust is only about 500 – 600° C; at the base of oceanic crust only 200 – 300° C. At this temperature, the Earth's mantle below the crust is still extremely stiff and by no means melted. Thus, the uppermost part of the mantle is welded to the base of the crust and is part of the lithospheric plates. Only much deeper down, at a depth where the mantle is about 1 200° C hot, does the mantle material become soft enough so that it begins to flow. Under the continents this is at about 120 – 200 km depth. Below this depth plate tectonics stop and the Earth's mantle convects in gigantic mushroom-shaped cells, not unlike boiling porridge. This part of Earth is called the *asthenosphere* (from asthenos = soft). As the oceanic lithosphere is denser and heavier than the continental lithosphere it lies mostly under water and vice versa. However, it is more or less coincidental that the total mass of water on Earth is just enough so that the major coastlines of this planet also form a geographic separation between continental- and oceanic lithosphere. In several regions on Earth oceanic lithosphere builds up land masses (e.g. in Cyprus or Oman) and there are also parts of the globe where continental lithosphere is under water (e.g. most parts of the Mediterranean Sea).

Oceanic plates are hard and stiff and drift around on the surface of the asthenosphere not unlike pack ice on arctic waters: the plates are formed at the mid-oceanic ridges and drift at several centimetres per year from there outwards. The further they drift away from the ridges, the colder they get and the more mantle material freezes to their base, thereby thickening the plate. This makes them denser and denser and they sink further and further into the asthenosphere. As a consequence, the water depth increases successively. At a distance of about 500 – 1 000 km from the mid-oceanic ridges, the plates are already so cold that they are denser and heavier than the asthenosphere below. They become gravitationally unstable and only float at the surface because of their stiffness, like a sheet of metal on a pond. In places where they collide with a continent, they quickly sink underneath them. Geologists call this *subduction*. The deep trenches before the coasts of Japan, Java, Sumatra, the west coast of South America or Alaska are such traces.

Continental plates are substantially softer, predominantly because they contain large amounts of radioactive elements that heat and soften the crust. As a consequence, continents deform much easier than oceans and build up the complicated folded mountain ranges of this planet. When they are pulled apart, they stretch like pizza dough and sedimentary basins from at their surface. One of the most famous sedimentary basins is the *Pannonian Basin* just east of the Alps.

The cycle between the stretching of a continent, its catastrophic tearing, the formation of new oceans, all the way to the subduction of these oceans and formation of a mountain range is an integral part of the history of the Alps and has dominated their evolution for the last 400 million years.

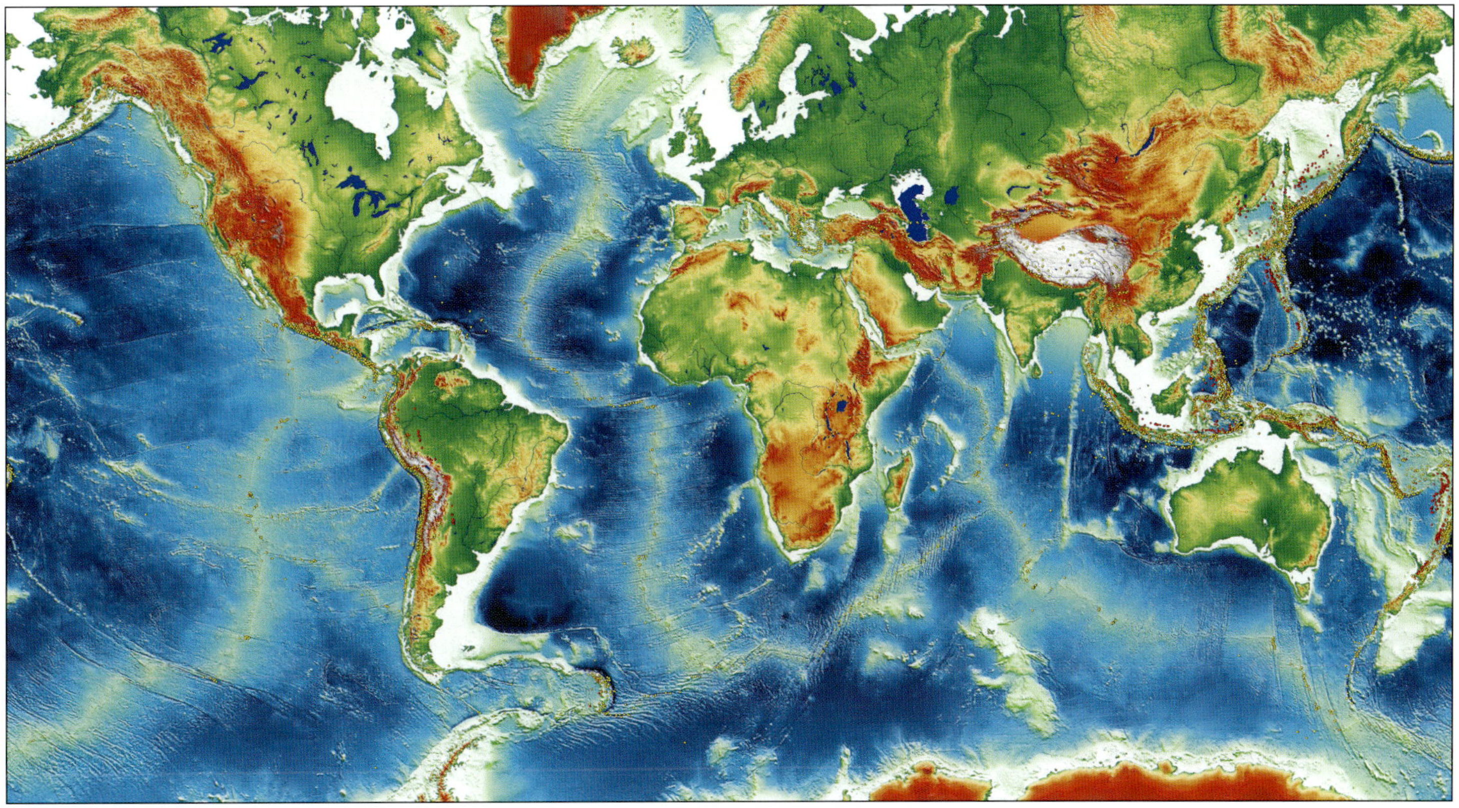

Topographic world map from 60° north to 60° south. The small dots show the earthquakes since 1970. The mid-ocean ridges are the narrow mountain ranges in the centre of the oceans. Where oceanic plates are pushed underneath a continent, deep sea trenches form, for example along the coast of Alaska or on the east side of South Georgia.

1.3. Subdivision of the Alps

From Nice to Vienna the Alps span about twelve degrees of longitude (5° – 17° E) and from Zurich to Milan some four degrees in latitude (44° – 48° N). As such, the range is about 1 000 kilometres long and 300 kilometres wide and is one of the smaller mountain ranges of this planet. However, the Alps are the best known and most densely populated mountain belt on Earth and have therefore been studied and subdivided with respect to many aspects. Geologically, is it possible to subdivide the range according to different criteria, for example: 1. *tectonically* (plate tectonic categorization) 2. *lithologically* (rock types) or 3. *palaeogeographically* (where the rocks were at the time of formation). For example, the granites forming the top of Mt. Blanc may be referred to as *"External Massif"* or simply as: *"granite"* or as: *"Variscan part of the European plate"*, depending on whether the tectonic position, the rock type or the age is meant. To the climber and friend of nature only the geographic subdivision is usually of interest:

Geographic Limits and Subdivision: The geographic limits of the Alps are simple: in the east they are bound by young sedimentary basins: the Vienna- and the Styrian Basin. To the north the Alps are bound by the Danube and – further west – by the large Bavarian lakes: Chiemsee, Starnberger See and Bodensee (Lake Constance). To the southeast the Alps merge into the Dinarides without a clear separation line. However, the Slovenian Pohorje Massif and the Julian Alps are definitely still part of the Alps. To the south the Alps are clearly bound by the Po plains and the boundary to the Apennines is defined by the Sestri-Voltaggio Zone just west of Genua. In the very west, the Alps are not very clearly delimited beyond the Rhône Valley but in the northwest, the Swiss Jura mountains are usually not considered to be part of the Alps.

In terms of a geographic subdivision, the Alps are divided into *Western Alps*, *Central Alps*, *Eastern Alps* and *Southern Alps*. The separation line between the Western Alps and the Central Alps leads from Lake Geneva up the Rhône Valley across the large St. Bernhard Pass and into the Aosta Valley. Therefore, Mt. Blanc is part of the Western Alps, but the famous mountains of the Bernese Oberland and the Valais Alps are part of the Central Alps. The boundary between the Central Alps and the Eastern Alps starts at Lake Constance and leads up the Rhine Valley across the Splügenpass to Lake Como. The border between the Southern and Eastern Alps is formed by a series of east-west striking valleys including the Gail Valley, the Puster Valley and the valley of the river Adda (Valtellina).

Tectonic Subdivision: Tectonically, the Alps are subdivided according to the geographic distribution of land masses in the Mesozoic: 1. The crystalline mid-crustal basement of the European plate; 2. the sedimentary cover of the pre-Mesozoic European plate; the region of the Penninic Oceans; 4. the Austro-Alpine and the South Alpine. The Austro-Alpine and South Alpine are both part of the former Adriatic plate.

In the Mesozoic some 200 million years ago, these four major units were arranged from north to south beside each other (p. 48). Today they are stacked on top of each other, but are still visible in their original relative position on transects from northwest to southeast (p. 53). The crystalline basement of the European plate is today only visible in the External Massifs of the Central Alps. The gneisses of the Aare Massif, the *Mt. Blanc Massif* or the *Belledonne Massif* are part of this. The sedimentary cover of this basement is to be seen in its original position in the Dauphinois and the Swiss Jura. In the Helvetica nappes of Switzerland, this sedimentary cover has been scraped off its crystalline base and displaced to the north. In the east, these nappes reach as far as the Allgäu Alps. The rocks of the Mesozoic Penninic oceans form many of the mountains of the Valais Alps and most of the Western Alps all the way to Genua. The Austro-Alpine starts in the Swiss canton Grisons and builds up the entire Alps east of this line. The South Alpine is separated from the Austro-Alpine domain by the same line that separates the Southern Alps and Eastern Alps. Until about 35 million years ago, South Alpine and Austro-Alpine were one single unit and thus the rock types are similar too.

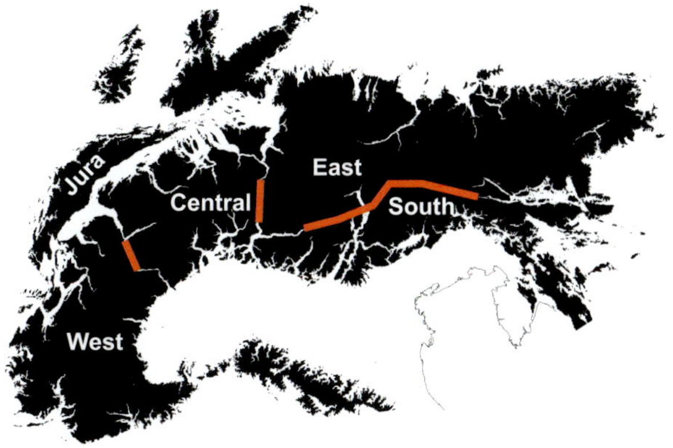

Geographic subdivision of the Alps. The region above 600 m sea level is shown in black. The red lines separate the Alps into Eastern-, Central-, Western- and Southern Alps.

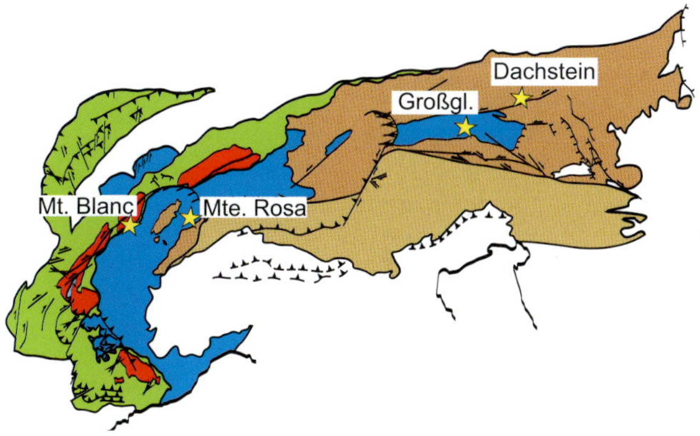

Tectonic subdivision of the Alps. Crystalline basement of the European plate = red; sedimentary cover and Helvetic nappes = green; rocks of the Penninic ocean floor = blue; rocks of the Adriatic plate = brown (dark brown = Austro-Alpine; light brown = South Alpine).

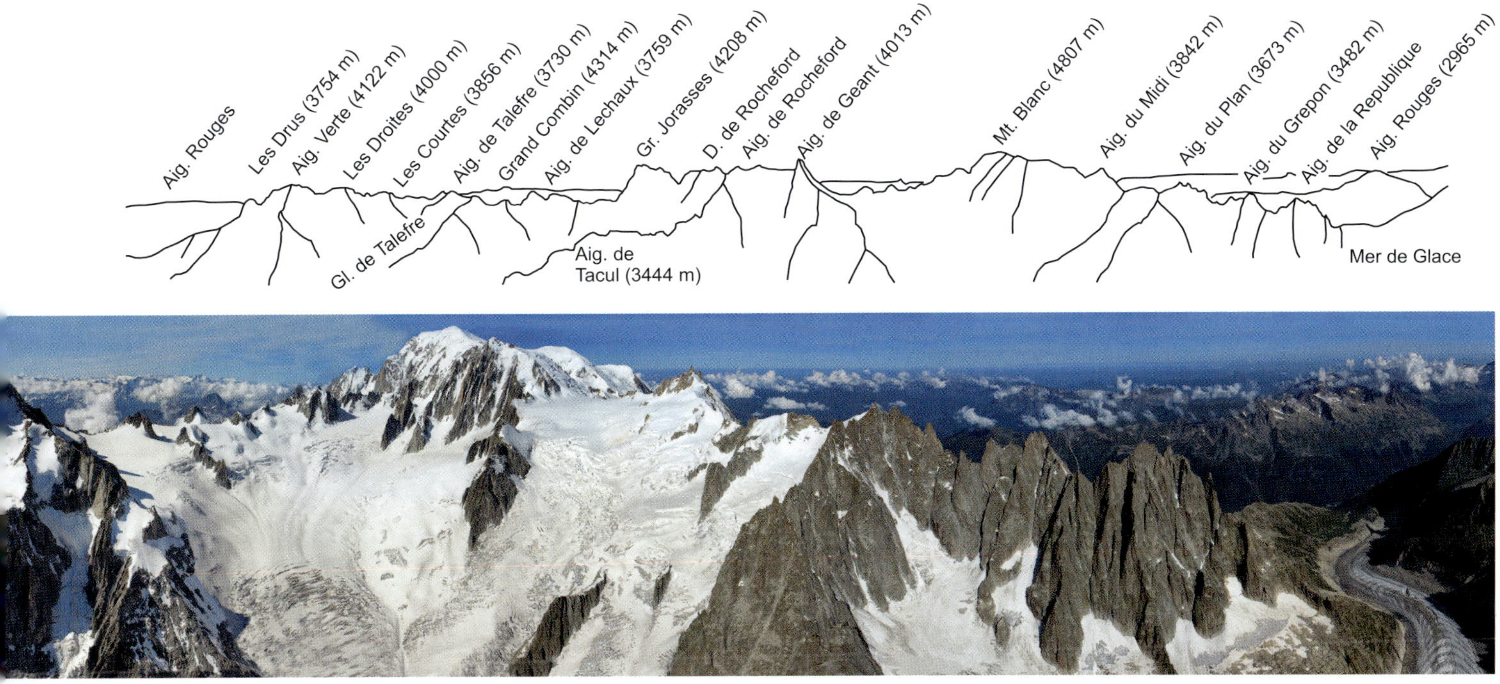

Left: 360° panoramic view of the Mt. Blanc Massif, taken from about 4000 m elevation above the Mer de Glace.

1.4. Elements of Mountain Building

Page 23: Deep down in the Val Verzasca of the Swiss Ticino region there are spectacular exposures of the widely folded rocks of the Lepontine Dome. The photo was taken in the cliff faces above the village Sonogno on the slopes of the Poncione d'Alnasca (2300 m), the: "El Capitan of the Alps". The white bands are deformed former granitic melts. The brown rocks in between are mica schists. Folding occurred at 600°C in about 30 km depth some 40 million years ago.

Right: Schematic three dimensional sketch of the folding of the Helvetic nappes of western Switzerland. It may be seen that the Morcles-, Diablerets- and Wildhorn nappes have come to rest on top of the Aiguilles Rouges Massif. They were once pushed in the Miocene into this position from the southeast (SE) towards the northwest (NW). The better part of the folded nappe stack has since been eroded away. The red ring shows the position of the fold on the next double page.

Page 24/25: The fold in the summit of Dent de Morcles (2969 m) high above the Rhône Valley near Martigny is one of the best known examples of folding of rocks in the Alps. The rock layers get younger and younger towards the base of the mountain, indicating that the entire pile of sedimentary rocks in Dent de Morcles lies upside down. The three dimensional sketch at right schematically illustrates the position of Dent de Morcles within the Helvetic nappes of Switzerland.

Mountain building processes are often subdivided into endogenic and exogenic forces. *Endogenic forces* are those that find their origin inside Earth and build up topography. *Folds, tectonic nappes* and *faults* are the observable record of these forces. *Exogenic forces* are those that destroy the topography from the outside: *rivers, landslides* and *glaciers* are the principal agents causing this destruction by erosion. The following 23 pages are dedicated to these six elements of mountain building.

A. Folds

Folded rocks are surely one of the most impressive records of the endogenic forces that act to build up mountains. They illustrate that rocks can indeed deform like honey. Part of earth science is exclusively involved with explanation of folds. In very general terms a rock is called folded when it bears a curved structure that is known to have been once linear. As such, curved bands in limestone, schist and gneisses are generally formed by folding, but bends in rivers or frozen lava flows are often not. *Ductile folding* is one of three fundamentally different mechanisms for how rocks can be deformed in the crust. The other two are *brittle fracturing* and *elastic bending*. The fact that rocks can break may be seen in every quarry. The elasticity of rocks is so small that it may only be seen on the scale of several hundreds of kilometres. However, folding may occur on all scales and in almost every rock type, so that folds are the most useful aid to help geologists to infer the evolution of a mountain belt.

In order to describe and explain folds, geologists discern between upright and flat-lying, symmetric and asymmetric, open and tight, shear, box, parasitic and bend folds, and many others. Many of these terms are used to describe the mechanics of folding. Others again are used to describe the geometry and orientation. Geologists also know terms for different parts of a single fold: fold limbs, fold axes, fold traces and fold hinges are only a small number of the terms used to describe folds. *Geological compasses* are used to measure the orientation of folds in rocks that are then used to infer transport directions of entire mountain building processes.

But how is it possible at all to fold rocks like yeast dough? Interestingly, this does not even require unusually high temperatures, pressures or large forces. Just as the daisy flower breaks through the concrete of an unused road, so rocks can fold if slow constant force is applied over long times. In general, pressures corresponding to depths between five and ten kilometres and temperatures around 200°C to 300°C are sufficient for rocks to be able to fold. A large part of the rocks of the Alps were buried at some stage of their geological history to at least this depth, so it comes as no real surprise that many rocks are folded.

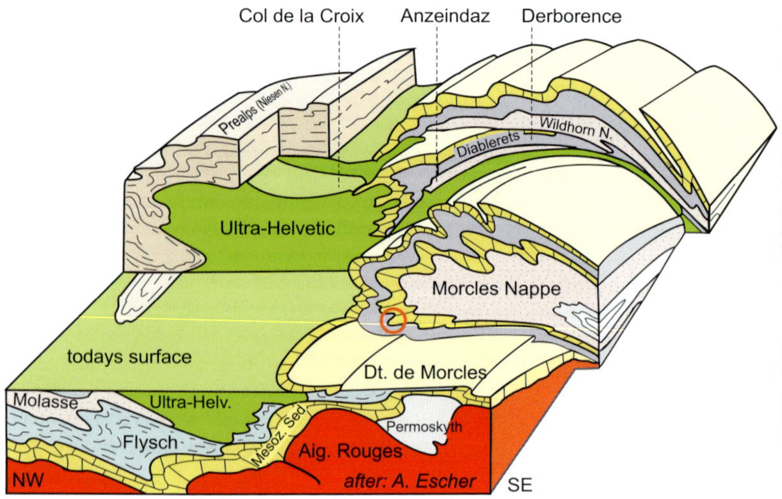

Page 27: The Tschingelhörner with the Martinsloch hole in the Glarus region of Switzerland seen from the north. The Glarus Thrust is the knife sharp horizontal line that separates that dark rocks above (Verrucano, 250 Ma old volcanic rock) from the light coloured rocks below (40 – 50 Ma old flysch sediments). The occurrence of older rocks on top of younger rocks has long caused puzzlement amongst geologists until it was realised that the different packets of rock are separated by a shear zone with many kilometres of displacement.

Right: A fault seen from the inside. In the Hirschgruben cave in the Hochschwab region of Styria in Austria (p. 142). The striations along a fault identify it as an extensional fault.

Bottom: A classic but false interpretation of the geology of the Glarus region of Switzerland from the time when the Glarus Thrust was not yet recognised as a fault zone. A. Heim suggested in the 19th century that two facing fold hinges between Gulderstock and Piz Sardona can explain the position of older rocks on top of younger rocks in the region. Today it is known that the Permian Verrucano and Eocene Flysch rocks were juxtaposed by many kilometres of horizontal displacement along a centimetre wide shear zone: the Glarus Thrust.

B. Faults and Shear Zones

Faults are discrete planar structures in rocks that separate packets of rocks that were displaced with respect to each other. Strictly spoken, only brittle fractures – as they are formed during earthquakes – are *faults*. More generally, geologists often also include *shear zones* in the term fault zone. Shear zones are narrow zones of ductile deformation. Faults, fault zones and shear zones form in all orientations and on all scales and with all kinds of displacements from millimetres to thousands of kilometres.

Steep faults with horizontal displacement are called *strike-slip zones*. In those, geologists discern between left lateral and right lateral strike slip zones, depending on whether one side of the fault is displaced with respect to the other to the left or right. The most famous right lateral strike slip zone on Earth is the *San Andreas Fault*, along which the Pacific plate moves towards the north relative to the North American plate. In the Alps, the largest fault zone is the *Periadriatic Line* (p. 28/29). However, most fault zones are very much smaller than these gigantic plate boundaries. Many faults are only few centimetres long and many break only individual crystals on a microscopic scale.

Inclined fractures at which the higher lying block is displaced upwards or downwards are called *thrusts* and *detachments*, respectively. Thrusts shorten the total length of both packets of rock and are a sign for horizontal compression. Detachments provide evidence for horizontal extension. Thrusting with substantial displacement causes uplifting of the upper block and lowering of the lower block. Detachments cause exposure of the lower block. Flat-lying fault zones like the *Glarus Thrust* can not be classified as thrusts or detachments. Faults and shear zones and their displacement are important evidence for the geologists to infer the stress field in the crust. Geochronological dating of minerals that have grown in shear zones is a well-established method to determine the age of fault zones.

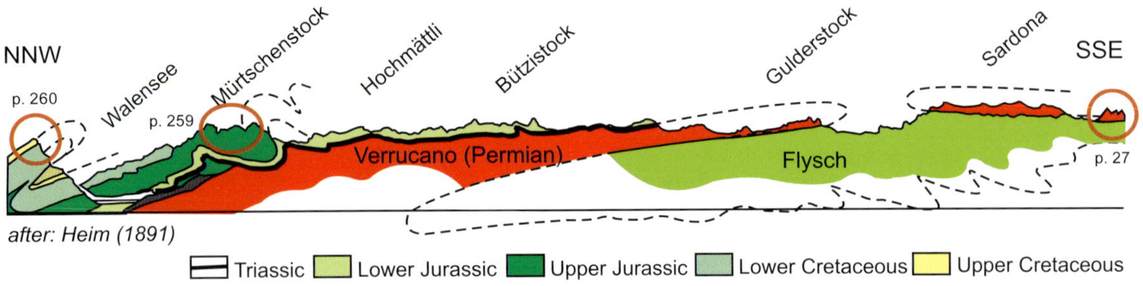

A garnet peridodite rock from the Alpe Arami near the north end of Lago Maggiore (photo: T. Nagel).

The Periadriatic Line is the largest fault zone of the Alps. It is almost 1 000 km long and several kilometres wide. In the Swiss canton Ticino it runs east – west along the *Magadino plain* just past the north end of Lago Maggiore. Because of the panoramic photo, the valley that formed along the fault zone appears curved. However, in reality, the Periadriatic Line separates the Eastern and Southern Alps along a straight west – east line (p. 110). The zone formed about 30 million years ago and separates the European plate from the Adriatic plate ever since. The right lateral displacement along the fault zone has deformed and crushed the rocks inside the zone so that rivers have cut their valleys along these soft zones. In its most active time, the motions along the fault zone have squeezed rocks from large depths to the surface. For example, on the *Alpe Arami* in the centre of the photo, a garnet peridodite was squeezed to the surface. *Gar-*

net peridodite is a rock type that normally only occurs in the Earth's mantle. The motions along the Periadriatic Line fault zone have ceased in the last few million years and the newly carved valleys were re-filled by sediments. The Magadino plain was only formed after the end of the last glaciation when sediments of the Ticino river re-filled the valley. Today it forms one of the few flat places in the Swiss canton of Ticino.

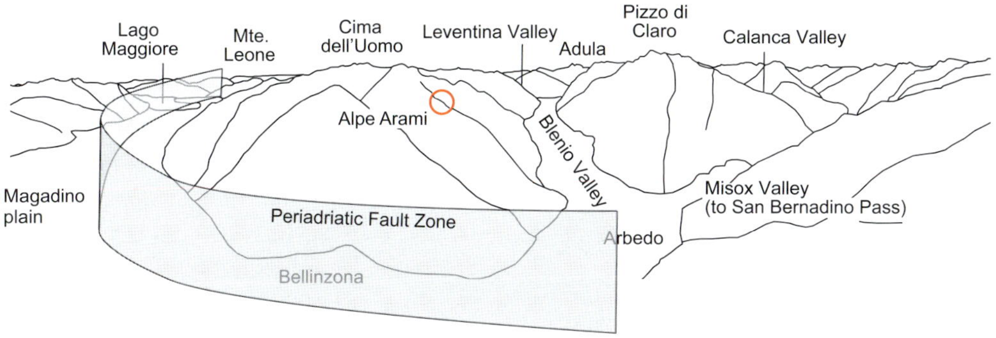

Right: The Helvetic nappes of western Switzerland seen from east to west (arrow on cartoon at base). Mt. Blanc and Dents du Midi are seen on the skyline.

Page 32/33: The Alpstein Massif with its highest peaks the Säntis (2501 m, left) and the Altmann (2435 m, right) forms the boundary between the Swiss cantons Appelzell and St. Gallen. The massif is the landmark of the Swiss city of Appenzell. The spectacular folds in the Mesozoic limestones of the massif are part of the Säntis-Drusberg Nappe. This nappe is part of the Helvetic nappe pile that also includes the Axen- and the Mürtschen Nappe in eastern Switzerland (p. 260). Lake Constance is in the left background. The geological cross section illustrates the position of the Säntis Nappe within the Helvetic nappe pile (colours explained on p. 258, ring = position of photo).

C. Nappes

The Alps are built up of *tectonic nappes*. Understanding the concept of tectonic nappes is fundamental to the modern understanding of the build up of many mountain belts. A tectonic nappe may be viewed as similar to a quilt. Just like a quilt, tectonic nappes have a large ratio of horizontal extent to thickness. Typically, tectonic nappes are only a few kilometres thick at most, but measure many tens of kilometres in horizontal extent. Nappes can be transported over large distances without losing their internal structure. Just like a quilt, they may be pulled, stacked, folded or shortened.

It was only about 100 years ago that the concept of nappe formation was first understood. Until the end of the 19th century the structure of the sedimentary rocks in the Alps was explained in terms of a continuously deposited sequence (*stratigraphy*). In layered rocks where older rocks were known to lie on top of younger rocks, large scale folds were inferred in order to explain how the sequence has been turned upside down. Many of the 19th century historic sketches of the geology of the Alps therefore include large folds in unexposed regions or in the air. However, even the most generous interpretation of folds was sometimes insufficient to explain the observed geology. Soon geologists learnt that large horizontal displacements between different nappes were necessary to explain what was seen.

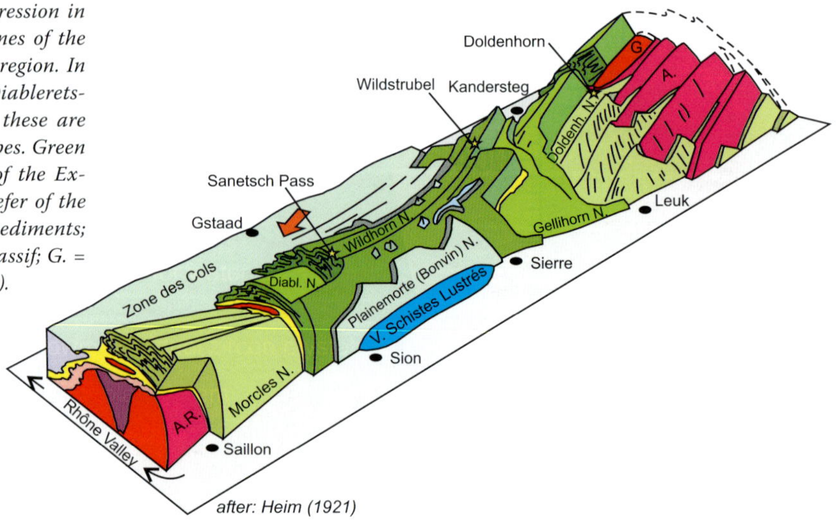

Bottom: Block diagram of the Helvetic nappes of western Switzerland. The nappe pile was uplifted the most in the very west and very east thereby forming the Rawil Depression in its centre. The nappes bear the names of the most important mountains in their region. In the west these are the Morcles-, Diablerets- and Wildhorn nappes. In the east these are the Doldenhorn- and Gellihorn nappes. Green = Helvetic nappes; red = gneisses of the External Massifs; blue = Bündnerschiefer of the Penninic realm; yellow = Neogene sediments; A.R. = Aiguille Rouges, A. = Aare Massif; G. = Gastern Massif (arrow = photo, p. 31).

The theory of nappes was born in connection with the interpretation of the geology of the Glarus region in eastern Switzerland. There, ancient volcanic rocks (Permian Verrucano) lie on top of much younger flysch rocks. Prior to the discovery of tectonic nappes, Albert Heim – one of the most famous Alpine geologists of the late 19th century – explained this geometric relationship by suggesting a double fold in the sky between Gulderstock and Piz Sardona (p. 26). With this "invisible fold" it was possible to suggest that the region is an upside down, layered stack. Subsequently, E. Suess and M. Bertrand independently discovered in 1883 and in 1884, respectively, that the Alps may be subdivided into tectonic nappes. Soon, Albert Heim's double fold interpretation became obsolete: it became clear that Verrucano and flysch rocks belong to different nappes which once were separated from each other by large distances. Only the later motion along the *Glarus Thrust* juxtaposed the rocks of different ages.

Legend says that Heim himself later accepted that his own interpretation was wrong, but he explained that his idea was only created for reasons of modesty. Is it not difficult to accept that nature itself can carry entire tectonic nappes over such large distances? In some ways he was right: even today, the mechanics of nappe formation is not very well understood.

Morcles
Diablerets
Wildhorn

North
(Lake Zurich east end)
Axen Nappe
(Glarus-) Verrucano
South
(Hinterrhein)
Säntis-Drusb. N.
Mürtschen N.
Sardona Flysch
G.T.
Glarus Nappe
Northhelvetic Flysch
Glarus
Thrust (G.T.)
Blattengrat
Flysch
Aare Massif
Gotthard
Tavetsch Massif
after: Schmid (1973)

D. Rivers

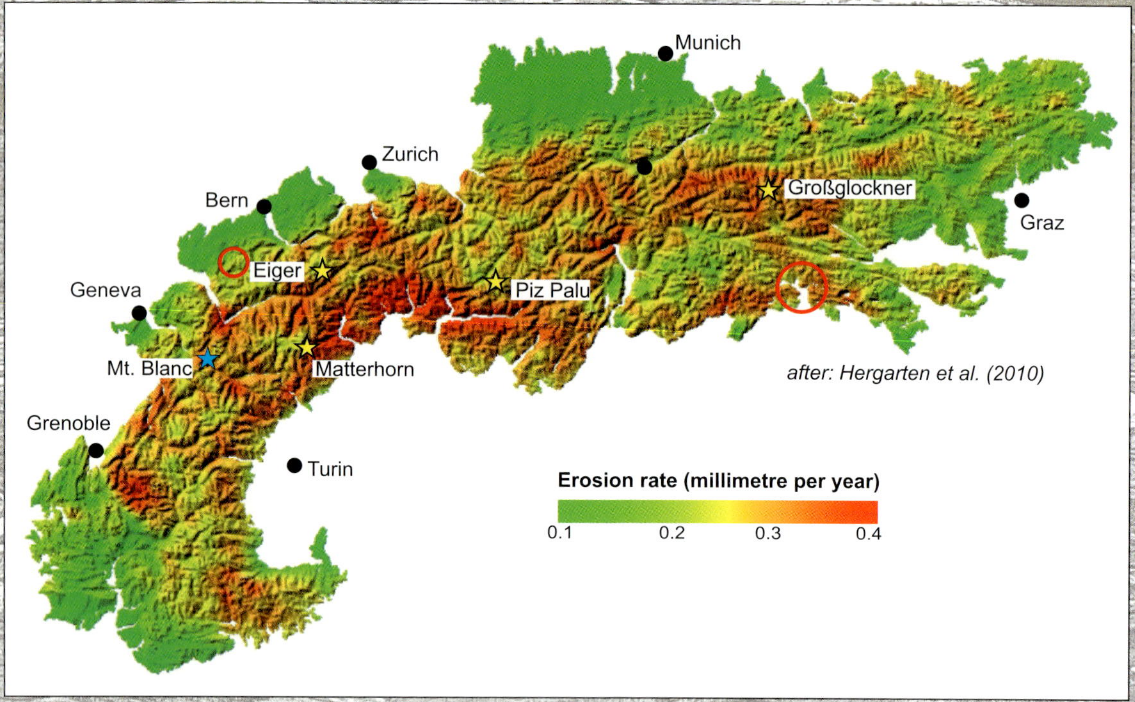

Geomorphologists know today that the erosive power of rivers depends largely on two factors: the amount of water and the topographic gradient. Both are factors that are usually quite well known. It is therefore possible to predict erosion rates using computer models. The map (above) shows the results of a computer model that was used to predict erosion rates from a digital elevation model (large ring = photo on p. 36; small ring = photo facing page).

Page 35: Streams on all scales erode the soft rocks of the Helvetic Wildhorn Nappe on the Sanetsch Pass south of Gstaad in Switzerland and produce a graphical fractal network.

The power of erosion destroys the Alps at about the same rate as the range is built up by plate tectonic forces. *Rivers* are the most important agent that transports eroded material out of the mountain belt. Other important agents that move material around are *landslides* and *glaciers*. However, landslides and glaciers usually transport material only on a local scale. In contrast, rivers transport material over large distances and continuously erode the Alps on all scales from the smallest mountain stream to large rivers like the Danube.

Four large river systems drain the Alps (p. 116): the Danube (101 000 km^2 catchment size above Vienna), the Po (80 000 km^2 at its mouth), the Rhône (100 500 km^2 at its mouth) and the Rhine (10 900 km^2 at Lake Constance). Together these four rivers transport roughly a hundred million tons of rocks from the Alps each year. This mass is divided up between three different types of river load: a small part is *dissolved* in water, a much larger part is the *suspended* load and another large part is transported as *gravel along the river bed*.

The rivers end up in three different oceans: the Danube in the *Black Sea*, the Po and the Rhône in the *Mediterranean*, and the Rhine drains into the *North Sea*. Five large sedimentary basins capture the sediment freight of the rivers: 1. the delta of the Danube in the Black Sea; 2. the Pannonian Basin east of the Alps; 3. the northern Molasse Basin; 4. the southern Molasse Basin and Po Delta between Torino and Venice; 5. the Rhône Delta and – to a small extent – the Rhine Graben.

The oldest sediments in most of these basins were deposited about 35 million years ago. Geologists believe that this age roughly corresponds to the first uplifting of the Earth's surface to form a small mountain range that started to deliver sediments to these basins. Since this time, several tenths of millimetres are removed from the surface of the Alps each year. Since the onset of uplift of the range, rivers have eroded many kilometre-thick layers of rock from the surface of the range. However, this does not mean that the Alps were many kilometres higher 30 million years ago: just as a submerged iceberg rises out of the water as it melts from above, so does the Earth's crust when it is unloaded from above. The uplifting of the Alps as a mountain range is the result of a complicated interaction of tectonics and erosion.

Page 36/37: The gravel fans spilled by the Tagliamento into the Po plains are large enough to be seen on satellite images of Europe. The Tagliamento drains the Southern Alps in the most seismically active part of the entire Alps: the Friuli region between the Dolomites and the Julian Alps (p. 119). Landslides caused by earthquakes in the Tagliamento (Canale-) Valley supply the material. On the photo, the river can be traced all the way to its mouth into the Adriatic Sea west of Venice.

Page 39 left: Is this a "before and after shot" of the last glaciation period? The left hand image shows the Mer de Glace at the base of Mt. Blanc. Many valleys in the Alps would have looked like this near the end of the last glaciation period some 10000 years ago. In the centre of the image the glacier (including its side moraines) is about one kilometre wide and several hundreds of metres thick, ...

Page 39 right: ... almost exactly as wide and deep as the Lauterbrunnental in the Bernese Oberland. The shape of the valley is a textbook example for a glacier-carved U-shaped valley. The peaks along the skyline are (from left to right): the Mittaghorn (3892 m), Grosshorn (3754 m), Breithorn (3780 m) and the Tschingelhorn (3562 m).

Below: Result of a computer model, developed by geologists at the ETH in Zurich to describe glacial erosion. The diagrams show the predicted evolution of glacial retreat in the region of the Aletsch glacier at four different time steps.

E. Glaciers

Erosion by glaciers is the second important exogenic force that works on the destruction of the Alps and competes with the tectonic uplift of the range. A complicated dynamic interplay between the accumulation region (where there is more snowfall in winter than there is melting in summer) and the ablation region (where the melting is more rapid than snow accumulation in the mean) causes the flow of glaciers under their own weight. In some glaciers the ice at their base is so cold, so that it is solidly frozen to the rocks below, and only the upper portion of the glacier flows. However, most glaciers flow on top of a water bed at a speed of many metres per year. During this motion, ice and embedded gravel grind up the underlying rocks. During a single glaciation period (roughly 10000 to 100000 years), several hundred metre-thick layers of rock can be removed. Polished rock pavements and gigantic moraine walls are the impressive record of this efficient erosion process. Glacier *cirques and lakes* are a typical landform of formerly glaciated landscapes. The large gravel fans spilled by the large rivers into the Alpine foreland were also deposited during the glaciation periods.

Today, only about 3600 km^2 of the Alps are covered by glaciers. This is roughly two percent of the surface of the range. The largest connected icefield of the Alps is the Aletsch-Jungfrau region and covers some 350 km^2. The largest glaciers of the Eastern Alps (where they are called "Kees" or "Ferner") are the Pasterze (20 km^2) at the base of Austria's highest mountain, the Großglockner, the Gepatschferner in the Ötztal Alps (18 km^2) and the Mittelbergferner (12 km^2), also in the Ötztal Alps.

The Aletsch glacier is the longest glacier in the Alps. It is 23 km long. It originates in the Jungfrau region of the Bernese Oberland and eventually drains through the Massa Gorge into the upper Rhône Valley. Each of its medial moraines can be traced back to the confluence of two side arms of the glacier in the upper part. The glacier contains about 30 billion tons of ice and, together with its tributaries, it covers an area of about 130 km^2. The Aletsch glacier is not only known for its enormous size, but also as an example for the rapid retreat of the Alpine glaciers: since the *Little Ice Age* around 1850, its surface has lowered by almost 100 m and the glacier has lost more than 2.5 kilometres in length. On a warm summer day the glacier loses more than 80 000 litres of its mass per second which flows as water through the Massa Gorge into the Rhône river. 18 000 years ago, the popular tourist walk along the ridge between Riederhorn and Eggishorn (foreground at left on the photo) was still covered by ice. Nevertheless, it's not all about shrinkage:

findings of tree trunks near the terminus show that the glacier was a lot smaller even 3000 years ago than today. Since then, the glacier has experienced several fluctuations that were the consequence of a complicated interplay of climate, ice dynamics and variations in the catchment contributions. Human impact on climate change is unquestionable today. However, it is important to remember that glacial fluctuations are sometimes a poor indicator for such climate change.

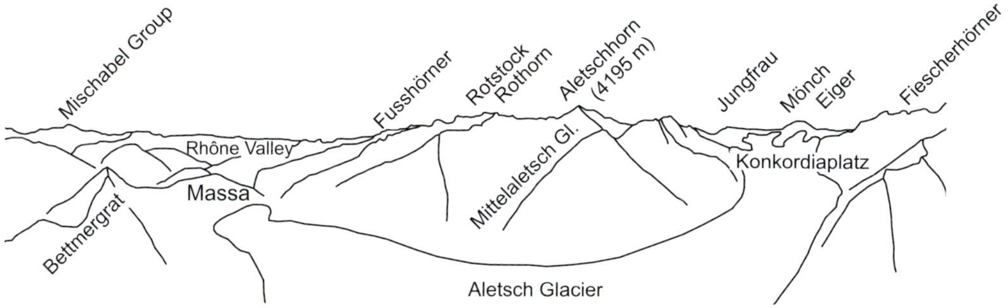

Page 44/45: The landslide of Flims is the largest landslide in the Alps and the second largest known landslide in the world. It involved more than 10 km³ of rock (in blue on the map at left) that slid into the Rhine Valley around 10 000 years ago. The landslide mass consists of limestone and marls of the Helvetic nappes. These rocks detached in 500 m thickness at an elevation of 2 700 m and skidded to the valley more than 2 000 m below. The rock mass piled up to 750 m high and formed a huge lake behind it: Lake Illanz. This lake remained for 1 000 to 2 000 years. Only then did the Rhine manage to carve a new valley through the rock mass. This valley is known as the Rhine- or Ruinaulta Gorge.

F. Mass Movements

Mass movements are the third important *exogenic force* that contributes to the destruction of the Alps as a mountain range. Mass movement is a term that is used for all erosion processes that remove material from the mountains outside the influence of streams. Mass movements include creeping slopes, landslides, mud flows, block glaciers, and catastrophic rock falls. Some of these mass movements move less than one metre per year, others with the speed of free fall. Mass movements are one of the few geological processes that occur at rates observable by man. Although mass movements account for only a small fraction of the total erosion of the Alps they are impressive documents of the rapid erosion that may occur and they are important suppliers of material that is then transported over long distances by rivers.

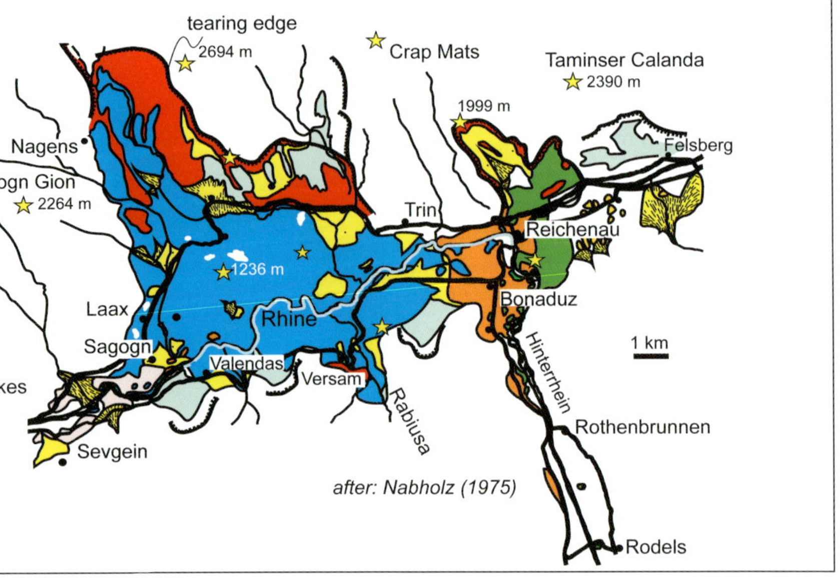

Geological map of the landslide of Flims (panoramic photo on next double page)

The heyday of landslides in the Alps was the time after the end of the last glaciation about 10 000 years ago. The retreating glaciers left deeply carved, oversteepened, instable U-shaped valleys in their wake, that suddenly lacked their counter-support by ice. The climate became warmer, vegetation increased and annual temperature fluctuations between summer and winter increased. As a consequence, a countless number of landslides occurred between 7 000 and 10 000 years ago. Usually, the landslides were triggered by earthquakes or particularly warm summers.

The largest prehistoric landslides of the Alps are the landslide of *Flims* (next page), the landslide of *Köfels* (p. 190) and *Tschirgant*. Each of these three landslides involved several cubic kilometres of falling rock. In historic times, we know of landslides in the Blenio Valley (p. 29), at Mt. Granier in Savoy, at the Königssee in Bavaria (p. 166) and on the Dobratsch in Carinthia (p. 120).

The largest land or rock slides in the last 50 years occurred on the Eiger in 2006 (500 000 cubic metres), on the Dent du Midi also in 2006 (about one million cubic metres), near Schwaz in Tyrol (150 000 cubic metres), an ice fall at the Mattmark Dam in Switzerland (500 000 cubic metres), a rock fall at the Bischofsmütze in the Austrian province of Salzburg in 1993 and finally the man-made catastrophe of Vajont in 1963 (p. 122).

Right: Medergen is a small settlement of "Walser people" at about 2 000 m elevation in a seemingly dangerous position below several block glaciers. Medergen lies high above Langwies near Arosa in the Swiss canton Grisons. Block glaciers typically contain just enough ice so that they can move at many centimetres per year. However, from the outside, they are almost indistinguishable from large piles of scree. Only when seen from the air their glacier like shape is visible.

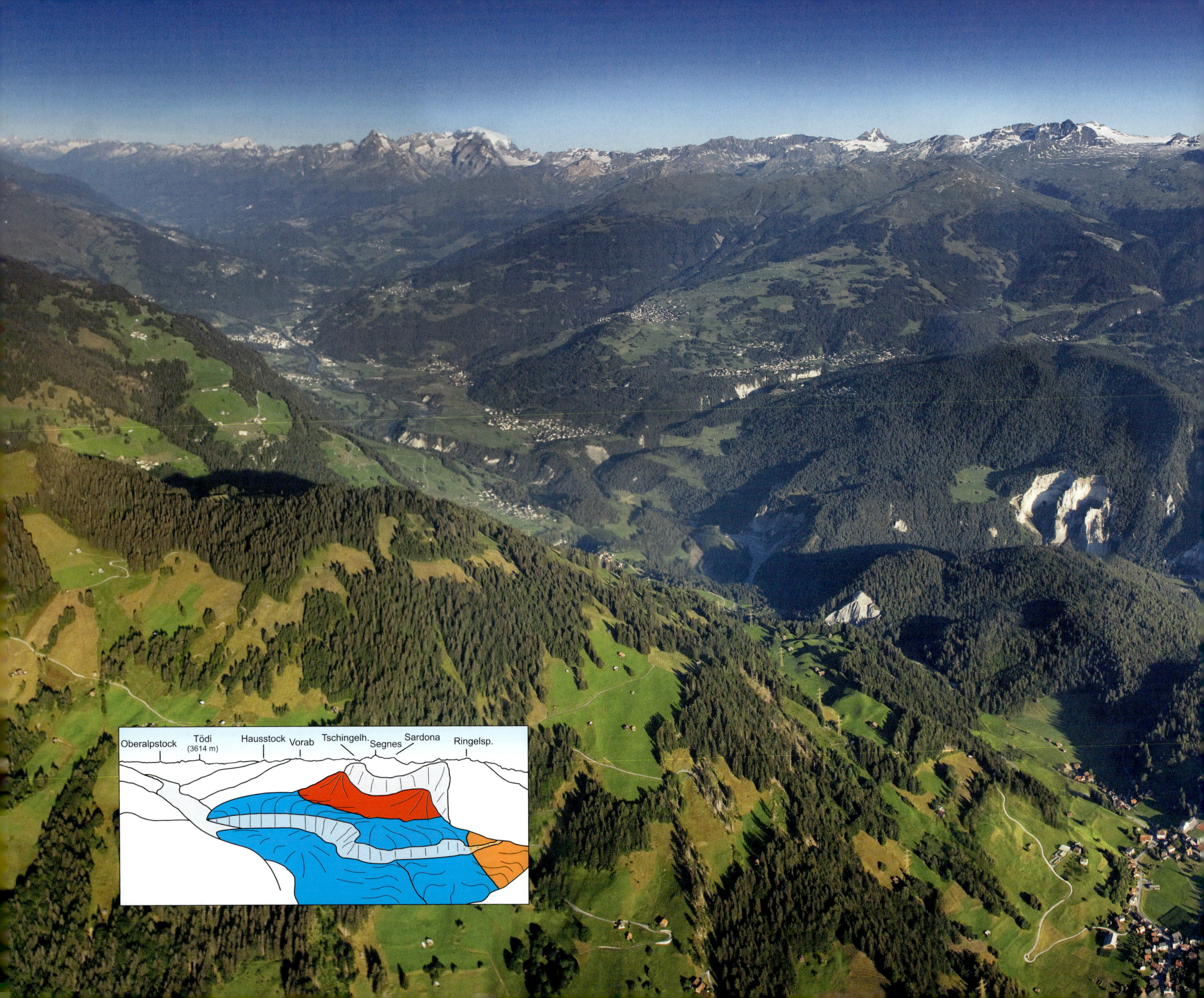
Oberalpstock
Tödi
(3614 m)
Hausstock
Vorab
Tschingelh.
Segnes
Sardona
Ringelsp.

2. Tectonic Building Blocks of the Alps

Austro-Alpine nappes (Adriatic Plate)

Northern Calcareous Alps & Greywacke Zone (Upper AA)

- 15 Juvavic nappes (Mesozoic cover)
- 16 Tirolian nappes (Mesozoic cover)
- 17 Bavarian nappes (Mesozoic cover)
- 18 Greywacke Zone (Paleozoic, strat. base of Tirol. nappes)

Upper Austro-Alpine basement nappes

- 19 Mesozoic cover of Upper Austro-Alpine basement nappes
- 20 Drauzug-Gurktal nappe system (Gurktal Alps, Graz Pal., Drauzug, Steinach nappe)
- 21 Ötztal-Bundschuh nappe system
- 22 Koralpe-Wölz high pressure nappe system (Schneebergerzug, Millstatt, Wölz, Saualpe, Koralpe)
- 23 Silvretta-Seckau nappe system (Silvretta, Innsbruck Quartzphyllite, Schladming, Seckau, Semmering)

Lower Austro-Alpine nappes

- 24 Lower Austro-Alpine nappes (Ela, Err-Bernina, Wechsel ...)
- 25 Nappes of the Sesia fragment (Dent Blanche nappe)

Penninic nappes

Upper Penninic nappes (Piemont-Liguria Ocean)

- 26 South Penninic ophiolites, Bündnerschiefer Préalps, Helminthoid Flysch

Middle Penninic nappes (Briançonnais Terrane)

- 27 Sedimentary cover of Middle Penninic basement nappes
- 28 Middle Penninic basement nappes
- 29 Detached Middle Penninic cover nappes
- 30 Permo-Carboniferous sediments (and their Mesozoic cover)

Lower Penninic nappes (Valais Ocean)

- 31 Tertiary flysch sealing Lower Penninic accretionary prism
- 32 North-Penninic ophiolites & Bündnerschiefer (including Rhenodanubian Flysch)

Parts of the European continental margin

Sub-Penninic nappes (distal European margin)

- 33 Mesozoic cover of Sub-Penninic basement nappes (including cover of the Gotthard Massif)
- 34 Non-eclogitic Sub-Penninic basement nappes (including Gotthard Massif)
- 35 Eclogitic Sub-Penninic basement units

Northern Alpine foreland and Helvetic nappes

- 36 Helvetic and Ultrahelvetic nappes
- 37 Helvetic Flysch
- 38 Subalpine Molasse
- 39 Deformed pre-Tertiary cover of the N. foreland (including Jura Mountains)
- 40 Undeformed pre-Tertiary cover of the N. foreland
- 41 External Massifs and Variscan basement

Tectonic Map of the Alps

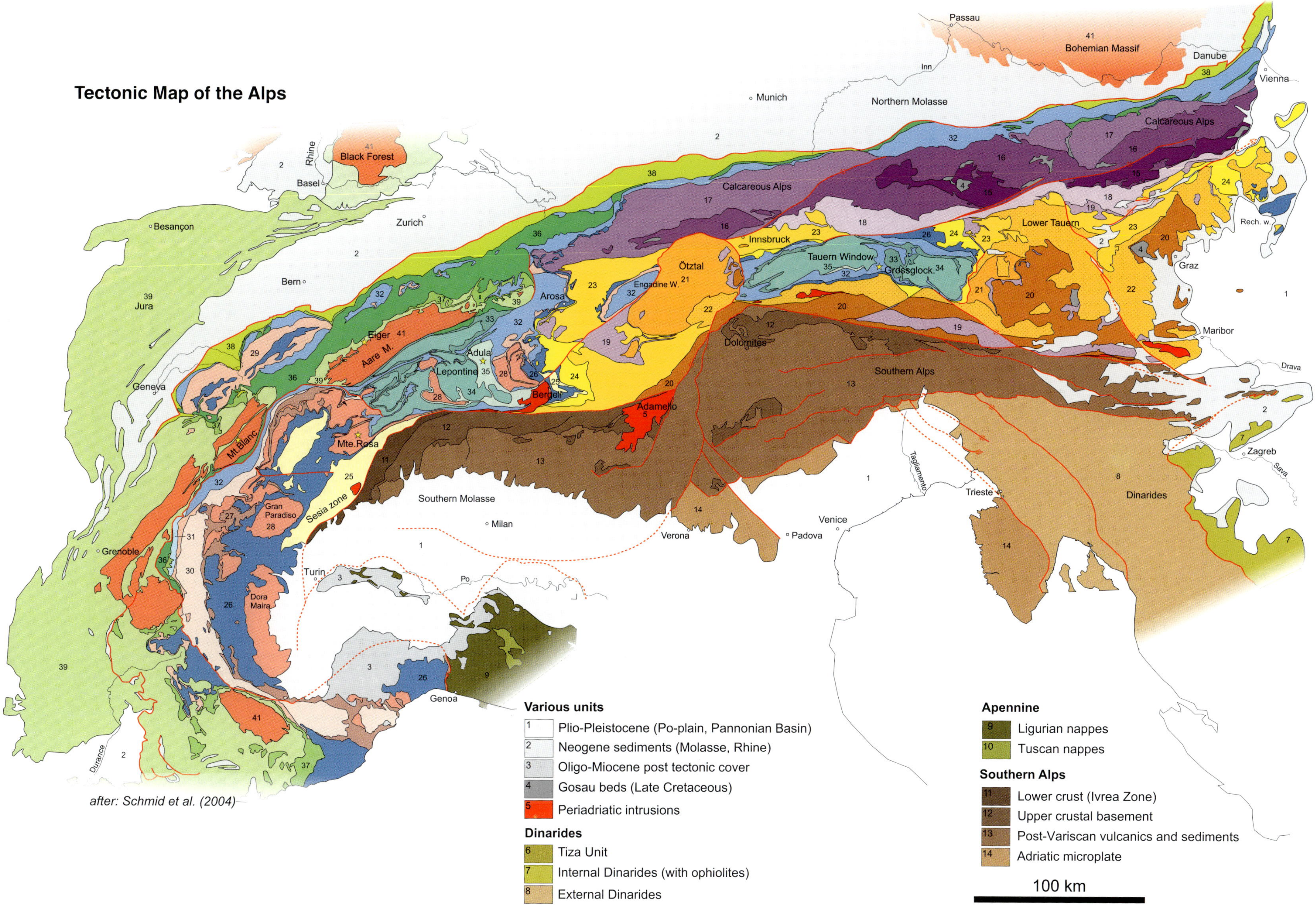

2.1. Palaeogeography und Subdivision

Schematic map of southern Europe and northern Africa in the Cretaceous (about 85 million years ago). The small yellow island in the Piemont-Ligurian Ocean is the Cervinia terrane. Todays position of the different plate fragments is shown on p. 21 and p. 53 (modified after Schmid et al. 2004).

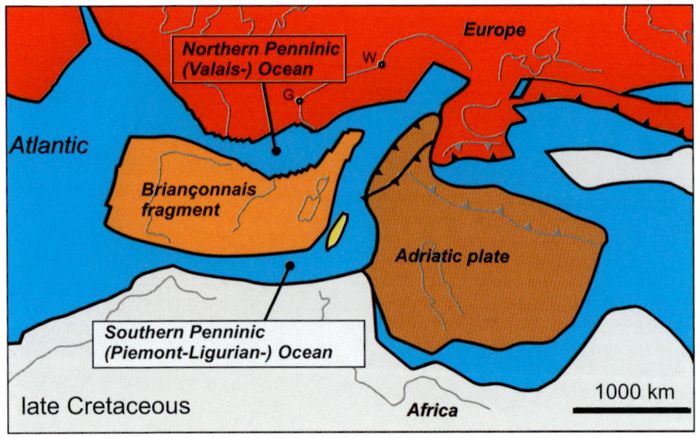

Tectonically, the Alps are typically subdivided on the basis of the geographic distribution of land masses in the Mesozoic (150 – 200 million years ago). At this time, several continental and oceanic plates covered the region of todays central Europe. These plates were later stacked on top of each other to form the Alps and some of them were completely pushed down into the Earth's mantle and disappeared. Not all plates grew or disappeared at the same time, but in a very schematic sequence from northwest to southeast, the following plates existed in the Mesozoic:

The European plate: The Mesozoic European plate once lay northwards of the Penninic Ocean and was connected to the ancient shields of northern Europe, not unlike todays understanding of Europe. The southern margin of this plate was intensely involved in Alpine tectonics.

The Northern Penninic (Valais) Ocean is the northern branch of the Penninic Oceans. It formed in the Jurassic and Cretaceous as the last part of an entire system of oceanic plates called the Penninic Oceans. Alpine orogeny later scraped off the upper parts of the crust from this ocean floor. These rocks are often called *Lower Penninic* because they have come to rest as the *lowermost* rocks in the tectonic pile of the Alps.

The Briançonnais Fragment formed some sort of peninsula or island in the middle of the system of Penninic Oceans. It formed the eastern continuation of the *Iberian peninsula* and was separated from the European plate by the Valais Ocean. The fragment is often referred to as the "middle continent".

The Southern Penninic (Piemont-Ligurian) Ocean is the southern branch of the Penninic Ocean system. Parts of the crust from this oceanic plate are often called the *Upper Penninic*, as they have come to rest in a relatively high position within the tectonic stack of the Alps. The ocean formed in the Jurassic as the easternmost branch of the Atlantic ocean. Amidst the oceanic rocks of the Upper Penninic, small splinters of continental rocks are found in the Central Alps. These splinters are interpreted as a Mesozoic island within this ocean that is called *Cervinia terrane*.

The Adriatic plate is often also called the *Apulian plate*. During the entire tectonic evolution of the Alps, this continental plate lay southeastward of the Penninic oceans. The rocks of this southern plate later became the Austro-Alpine and South Alpine and form the tectonically highest rocks in the Alps.

The Meliata-Hallstatt Ocean was an ocean that had formed on the south side of the Adriatic plate in the Permian. Today its rocks are only found in the Dinarides and the Carpathians. In the Alpine realm, this ocean was completely subducted into the mantle or laterally displaced eastwards by gigantic strike slip zones, so that it has not played a role in the evolution of the Alps since the Jurassic.

Page 49: The contact between the Adriatic continental plate and the Briançonnais Fragment at Schwarzhorn in the Rätikon region near the Austrian-Swiss border. The floor of the southern branch of the Penninic Ocean (Piemont-Ligurian Ocean) that once separated the two plates is here tectonically shortened to a mere few metres: a tiny splinter of the dark gabbros of this ocean floor (blue) remains between the rocks of the neighbouring continents. Colouring of the sketch corresponds to the map at top left. Green are flysch rocks that once were sediments on the Penninic ocean floor.

Drusenfluh
Schesaplana
Schwarzhorn

The famous ski resort of Arosa lies in the Swiss canton of Grisons on the geological boundary between the Austro-Alpine rocks (that belonged to the Adriatic plate in the Mesozoic) and the rocks of the Penninic Ocean floor. The so-called "Arosa-Schuppenzone" (green) is a mixture of deep-sea sediments (radiolarites), shallow water sediments (limestones) and continental rocks (schists and gneisses) along this boundary. The colouring of the sketch corresponds to the map on the previous page.

Although the tectonic subdivision of the Alps is based on the palaeogeography of Europe in the *Mesozoic*, the different units are now stacked up vertically above each other. Often they do not lie exactly vertically over each other, but are separated by inclined contacts. Geologists speak of *structurally higher* and *structurally lower*. The following tectonic parts are distinguished in the Alps with respect to their present position:

The European basement forms much of the crystalline base of the Alps. It consists of granites and gneisses that formed some 350 Ma ago during the *Variscan orogeny*. This basement dips southwards and disappears underneath the Alps along the northern flanks of the range. It forms the lowermost unit. However, this basement reaches the surface in places where it was domed up by tectonic processes. The *External Massifs* of the Central Alps (e.g. Aare- or Mt. Blanc Massif) and the *gneisses of the Tauern Window* in the Eastern Alps are such basement domes that belong to the European plate.

The sedimentary cover of the Mesozoic European plate is made up of shallow water sediments like marls and limestones. Parts of this sedimentary cover are still in contact with their base, for example in the *Dauphiné* or the *Swiss Jura* mountains. Other parts are completely scraped off their base and displaced many kilometres to the north. These are the Helvetic nappes of Switzerland.

The rocks of the Penninic domain consist of basalt, gabbro and sediments from the Penninic Ocean floor. They also contain fragments of continents that once lay along the margins or inside this ocean. Rocks are divided into *northern (lower) Penninic rocks* of the former *Valais Ocean*, continental rocks of the *Briançonnais Fragment*, oceanic rocks of the *southern (upper) Penninic (Piemont-Ligurian) Ocean floor* and the *Penninic rocks of the Eastern Alps*. Rocks of the Penninic domain build up much of the Central Alps and the rocks inside the Tauern Window of the Eastern Alps.

Rocks of the Adriatic plate constitute the Austro-Alpine and the South Alpine rocks. Austro-Alpine rocks form a colourful mixture of schists, gneisses, limestones and other rocks of continental origin. Until about 30 Ma, the Austro-Alpine continued directly to the south into the South Alpine and the South Alpine rocks are therefore built up of a similar mixture of rocks. However, since about 30 Ma the South Alpine has not been pushed north anymore, but south over the top of the Po plains. The South Alpine includes all rocks south of the Periadriatic Line, while the Austro-Alpine builds up most rocks of the Eastern Alps.

Sketches of the palaeogeography of Central Europe at three different time steps in the Mesozoic (after: R. Schuster).

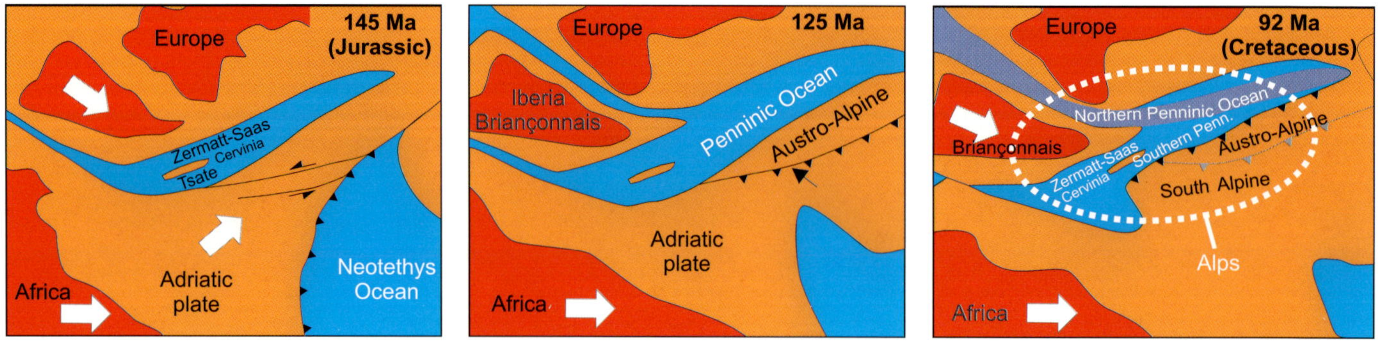

Tectonic block diagram of the entire Alps seen from west to east. The diagram illustrates how the different building blocks of the range lie on top of each other. From bottom to top these are: 1. The European basement (red). 2. The sedimentary cover of the Mesozoic European plate (green). 3. The domain of the Penninic Oceans (blue, and including oceanic rocks of the northern and southern Penninic Oceans, as well as rocks of the Briançonnais Fragment and Sub-Penninic units). 4. Parts of the Adriatic plate (dark brown = Austro-Alpine and light brown = South Alpine). 5. Neogene sediments of the Molasse basins are the uppermost and youngest unit (yellow). Intramontane basins in light yellow. Periadriatic intrusive rocks in pink. However, these uppermost rocks are currently being overthrust by parts of the Alps.

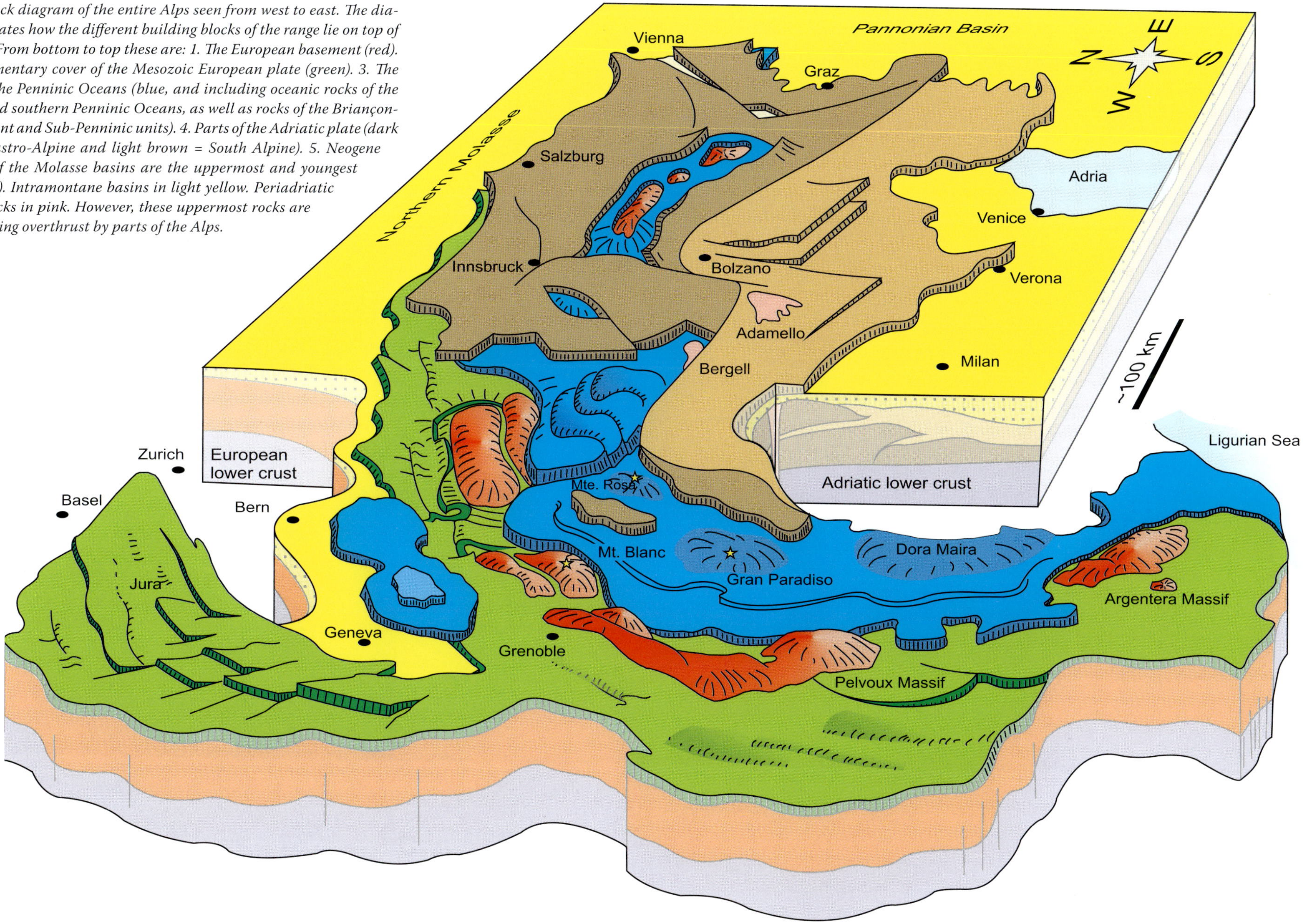

2.2. European Continent

A. The Base

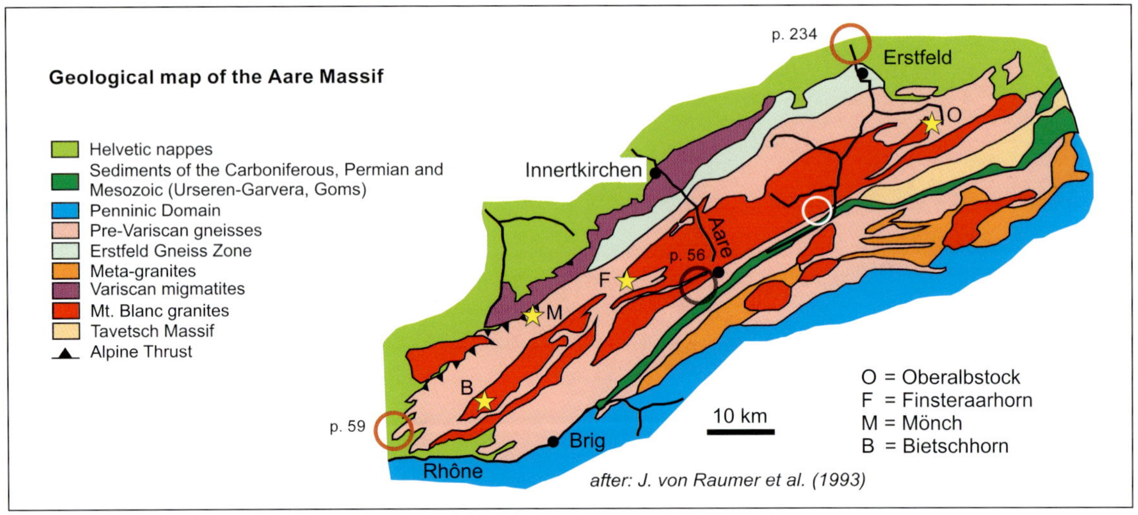

The deep base of the Mesozoic European plate is only visible in the *External Massifs* and the *Tauern Window*. These places are huge domes of the European plate where erosion has removed the higher lying rocks exposing the crystalline basement in a window-like fashion. From south to north five External Massifs are known: the *Argentera Massif*, the *Pelvoux Massif*, the *Belledonne Massif*, the *Mt. Blanc- / Aiguilles Rouges Massif* and the *Aare Massif*. The names for most of these massifs stem from their highest peaks: Argentera (3 292 m) lies south of Cuneo in the Italian Alpes Maritimes, Mont Pelvoux (3 932 m) in the Dauphiné-Alps of France gives the Pelvoux Massif its name, the chain of the Belledonne mountains northeast of Grenoble names the Belledonne Massif, the Aiguilles Rouges- and Mt. Blanc-Massif are named after the mountains of that name and the Aare Massif is named after the river Aare. Some smaller gneiss complexes like the *Tavetsch-Massif* south of the Aare Massif are also counted as part of the External Massifs.

A unifying feature of all External Massifs is that they disappear on all sides underneath sedimentary cover. At depth, the External Massifs are connected to the crystalline base of the European plate. The *Sub-Penninic nappes* are related to the External Massifs. However, in contrast, they are not connected with their base anymore (p. 62).

All External Massifs consist of granite, amphibolite, schist and high grade metamorphic rocks that formed between 280 and 300 million years ago in the European plate during the Variscan mountain building processes. Occasionally, traces of even older rocks are found. During the Alpine mountain building processes, the rocks of the External Massifs were re-heated to about 300° C, before they were domed up and pushed towards the surface. The fact that many of the highest peaks of the Alps lie inside the External Massifs is, however, only partly related to this doming. In addition, it has to do with the erosive resistivity of the rocks.

Right: Grandes Jorasses (at left), Dent de Geant (at right) and Aiguille du Midi (foreground at right) are some of the highest peaks of the Mt. Blanc Massif, one of the five External Massifs of the Alps.

Page 55: The Salbitschijen (2 981 m) in the central Aare Massif dominates the Göscheneralp valley and is one of the most famous climbing peaks of the Alps (white ring on the map above). The mountain is made up of 300 million year old granite. The map shows the five External Massifs of the Alps in red. Black Forest and Vosges are not part of the Alps but are geologically equivalent to the External Massifs (see also p. 67). Neogene sediments in yellow.

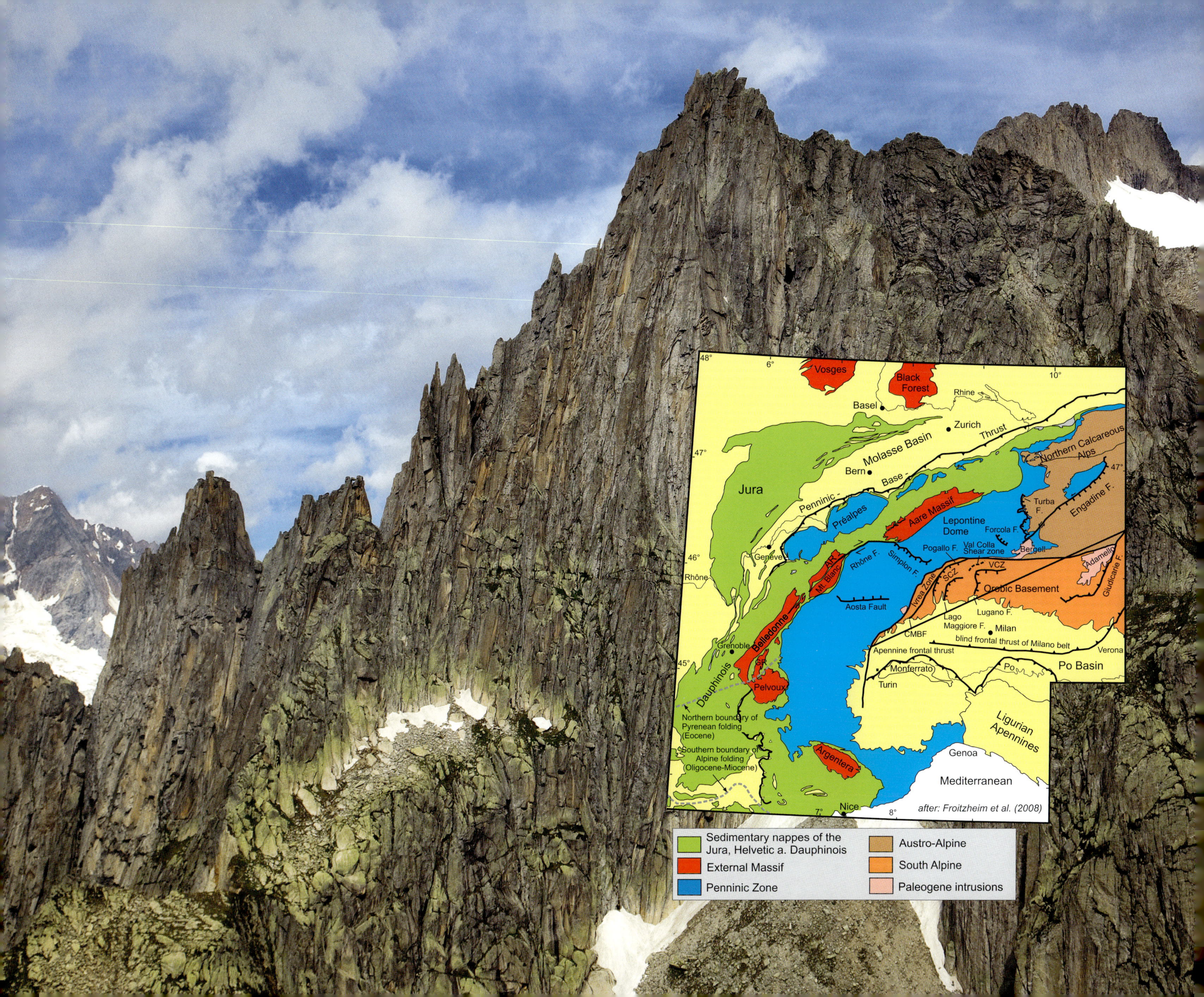

48°
6°
10°
47°
46°
45°
7°
8°
Vosges
Black Forest
Rhine
Basel
Zurich
Molasse Basin
Thrust
Bern
Base
Penninic
Jura
Préalpes
Aare Massif
Lepontine Dome
Northern Calcareous Alps
Turba F.
Engadine F.
Forcola F.
Val Colla Shear zone
Pogallo F.
Bergell
Adamello
Giudicarie F.
Genève
Rhône
Rhône F.
Simplon F.
AR
Mt. Blanc
Aosta Fault
Ivrea Zone
SCZ
VCZ
Orobic Basement
Lugano F.
Lago Maggiore F.
Milan
CMBF
blind frontal thrust of Milano belt
Verona
Apennine frontal thrust
Po Basin
Po
Monferrato
Turin
Grenoble
Belledonne
GR
Pelvoux
Dauphinois
Northern boundary of Pyrenean folding (Eocene)
Southern boundary of Alpine folding (Oligocene-Miocene)
Argentera
Ligurian Apennines
Genoa
Mediterranean
Nice
after: Froitzheim et al. (2008)
Sedimentary nappes of the Jura, Helvetic a. Dauphinois
External Massif
Penninic Zone
Austro-Alpine
South Alpine
Paleogene intrusions

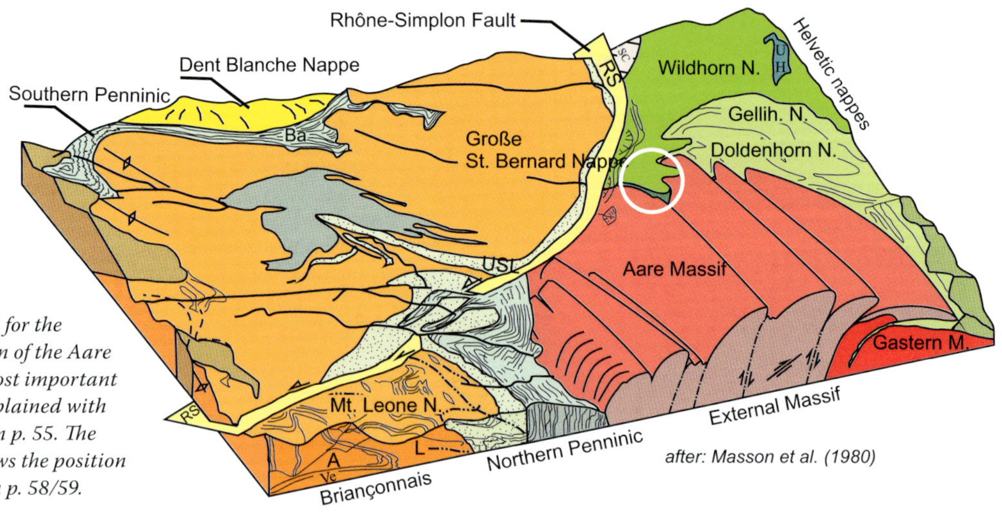

Block diagram for the western margin of the Aare Massif. The most important colours are explained with the diagram on p. 55. The white ring shows the position of the photo on p. 58/59.

The largest and easternmost External Massif of the Alps is the Aare Massif. It lies in Switzerland and stretches over about 100 kilometres length and 20 – 30 km width between the Lötschen Pass in the west and the Tödi region of the Glarus Alps to the east. It is named after the river Aare and many of the peaks of the Bernese Oberland are part of it. The Aare itself is the longest river that lies entirely within Switzerland. It has a catchment of 18 000 km^2. It is also the largest tributary of the Rhine. The Aare river begins at the Oberaare- und Unteraare glaciers in the Bernese Oberland. It then flows through the Grimselsee and then northwards through the Haslital towards Bern. The most important summits in the massif are the *Bietschhorn, Aletschhorn, Finsteraarhorn, Schreckhorn, Dammastock, Birsten* and *Oberalpstock* as well

as *Mönch* and *Jungfrau*. The famous Eiger is not part of the massif as it belongs to the Helvetic cover units. The rocks of the Aare Massif are gneisses and granites that formed in the Variscan mountain building process more than 300 million years ago. During Alpine tectonics, the Aare Massif was barely altered and eventually pushed to the surface less than 20 million years ago.

Directly to the south is the Gotthard Massif, which consists of similar rocks. However, the Gotthard Massif is often attributed to the Sub-Penninic units and not to one of the External Massifs. In the last few years, both the Aare and the Gotthard massifs were completely drilled through by the Gotthard Base Tunnel project (p. 234).

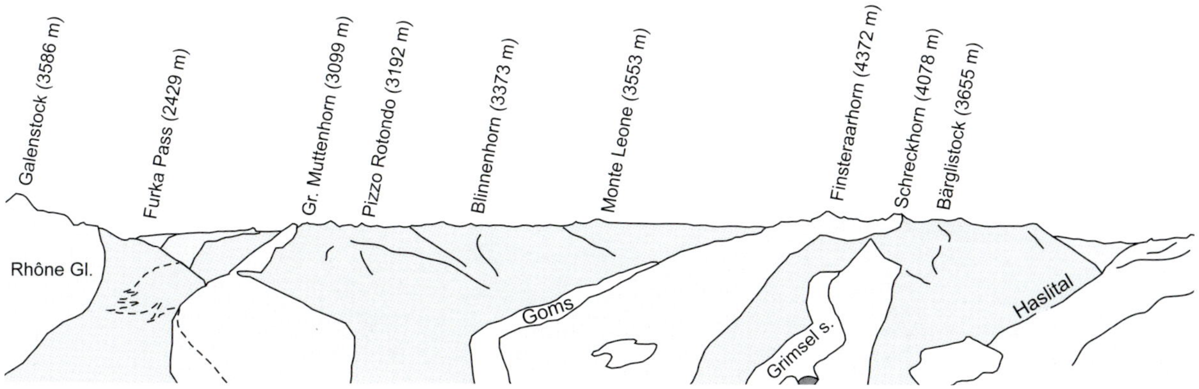

Both photos this double page: The intensely folded front of the Doldenhorn Nappe at the Lötschen Pass illustrates impressively how the Helvetic nappes (green in the sketch) were once transported northwards (from left to right) over the top of the Aare Massif (red) where they have come to rest.

Page 60/61: On the Bifertenstock (3419 m) east of Tödi mountain in Grisons, the sedimentary cover of Mesozoic Europe is still in direct contact with its crystalline basement. The deeply incised valleys of the Linth to the north (under the clouds at left) and the Rhine Valley (at right) uncover the rock layers all the way down into the gneisses of the Aare Massif. The apparent folding in the rocks in the foreground is only a sectioning effect: glacial carving of the cirques has caused the apparent curvature in the relatively flat layers (after the geological map of Bifertenstock of Keller, 1910).

B. The Sedimentary Cover

After the end of the Variscan mountain building processes about 300 million years ago, Europe was slowly covered by sediments deposited on to the crystalline basement. This process lasted throughout the Mesozoic. The sedimentary pile that developed is often very generally called: *"the sedimentary cover of Europe"*. Sedimentation started in the Carboniferous and Permian with widespread deposition of river sediments and volcanic rocks that filled the fractures and valleys of the hilly landscape of the European shield (e.g. Glarus Verrucano). In the Triassic, the continent sank to below sea level and the deposition of sediments continued with coral reefs and other shallow water sediments. This sedimentation lasted for almost 200 million years until the Eocene around 30–40 Ma. In the *Dauphiné* and in the *Swiss Jura mountains*, this sedimentary pile is preserved largely undisrupted. Around the margins of the External Massifs the way in which these sediments still lie in contact with their base is also visible. However, in large parts of the Alps the tectonic processes that formed the Alps scraped these Mesozoic sediments off their base and displaced them many kilometres to the north. These are the *Helvetic nappes of Switzerland,* which are known for their spectacularly folded rocks. In eastern Switzerland, the Helvetic nappes are (from bottom to top) the *Glarus Nappe, Mürtschen Nappe, Axen Nappe* and *Säntis-Drusberg Nappe,* named after mountains of the same name. In western Switzerland these nappes lie on top of the eastern margin of the Mt. Blanc- and Aiguilles Rouges Massifs and the western margin of the Aare Massif. East and north of the Rhône Valley these nappes are called *Morcles* and *Diablerets Nappe.* In the region of the western Aare Massif these nappes are called the *Doldenhorn-* and *Gellihorn Nappe.* The *Wildhorn Nappe* lies on top (p. 30).

Grand Combin
Mt. Blanc
Balmhorn
Hockenhorn
(3293 m)
Photo p. 58
Lötschen Pass

Nummulite limestone (Eocene)
Seewer limestone (Upper Cretaceous)
Gault (Upper Cretaceous)
Schrattenkalk limestone (Cretaceous)
Drusberg layers (Middle Cretaceous)
Siliceous limestone (Middle Cretaceous)
Valanginian limestone (Lower Cretaceous)
Öhrli limestone (Lower Cretaceous)
Malm limestone (Upper Jurassic)
Fe-rich sandstone (Oolitic) (Lower Jur.)
Triassic (Quartenschiefer, Dolomite)
Aare Massif (Tödi Granite)

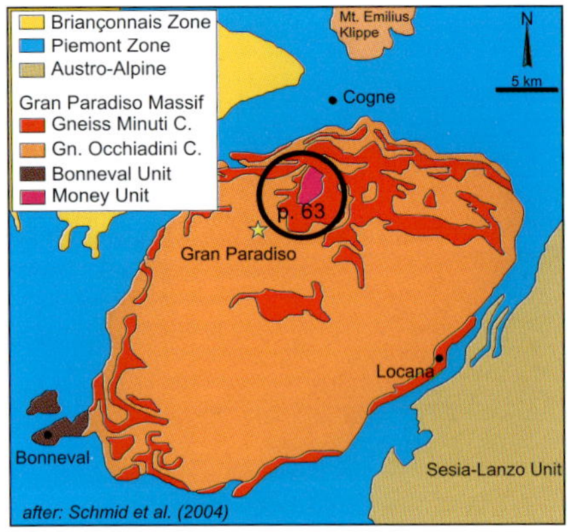

Geological map (top) and cross section (below) of the Gran Paradiso Massif. The cross section illustrates how the massif lies like a huge bubble surrounded by oceanic rocks in the Earth's crust. Legend for the cross section on p. 258.

C. The Continental Margin

The southernmost margin of Europe as it was in the Mesozoic formed the northern boundary of the Penninic Ocean. During the later subduction of this ocean, this margin was incorporated into the tectonic processes of the Alps and some parts were completely detached from the underlying Earth's mantle, deformed, heated and metamorphically changed. These parts are often only recognisable as parts of the former European margin because their chemistry identifies them as rocks of continental origin: quartz-rich gneisses, schists and granites. The northernmost part of this continental margin is the *Gotthard Massif* which is still connected to the Aare Massif at depth. South of the Gotthard Massif these parts are subdivided into two different groups: the *Sub-Penninic* units and the *Under-Penninic* units.

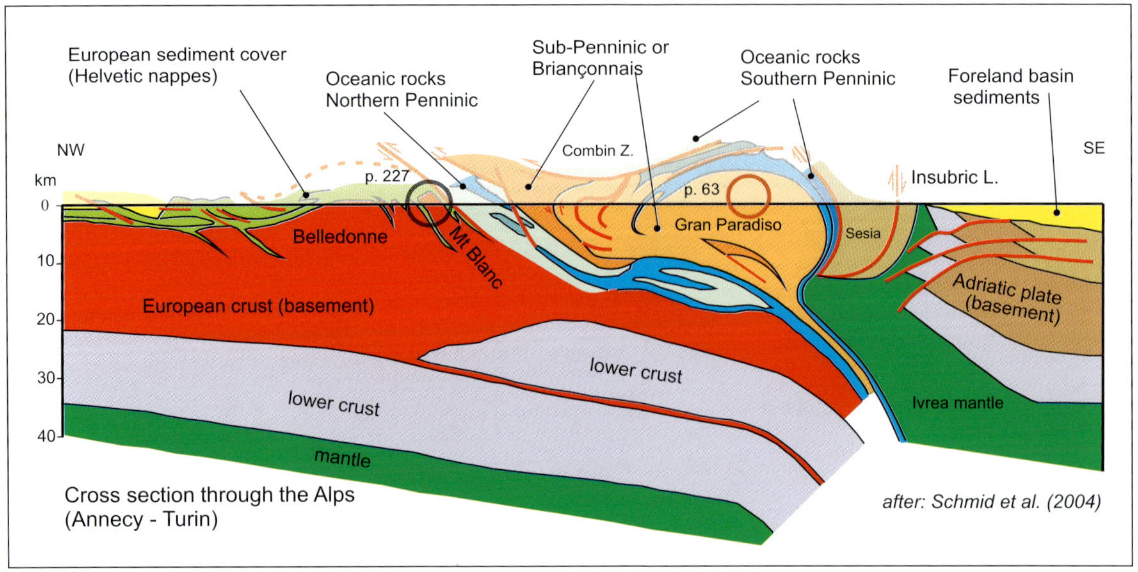

The *Under-Penninic units* involve the so-called *Internal Massifs*: the *Dora-Maira Massif, Gran Paradiso Massif* and the *Monte Rosa Massif.* All these massifs lie as gigantic drop shaped bubbles between the rocks of the former Penninic ocean floor. During Alpine tectonics they were metamorphically altered at extremely high pressures (eclogite facies). Because the massifs are not connected to their former base, their origin is often not easily identified. In the past, the Internal Massifs have been variably interpreted as parts of the European continental margin or as parts of the *Briançonnais Fragment.*

The classic *Sub-Penninic units* lie somewhat east of the Under-Penninic units: they include the gneisses of the *Lepontine Dome.* This huge, dome-shaped structure lies between the Rhône Valley to the north, Lago Maggiore to the south, Simplon Pass to the west and Misox Valley to the east. Sub-Penninic nappes in this region include the Simano-, Leventina-, Maggia-, Antigorio- and Monte Leone Nappes. The highest metamorphic grade of the rocks is reached in the south, where they include eclogites and garnet-bearing gneisses. Along the eastern margin of the Lepontine Dome, the Under-Penninic Adula Nappe lies on top of the Sub-Penninic units. The Adula Nappe also contains eclogites.

In the *Eastern Alps*, Sub-Penninic rocks are only visible inside the Tauern Window. There, erosion has removed enough of the overlying rocks from the former Penninic Ocean floor so that the Sub-Penninic *Zentralgneis* has become exposed. The gneisses of the Sonnblick, the Hochalmspitze, the Großvenediger and the Granatspitze are part of this (p. 158).

Page 63: Gran Paradiso and La Grivola seen from the north.

Punta Ceresole (3777 m)
T. di Valnontey (3642 m)
Gran Paradiso (4061 m)
B. di Montandayné (3838 m)
Punta Rossa (3630 m)
Grivoletta (3514 m)
La Grivola (3969 m)
Cresta Della Forcia

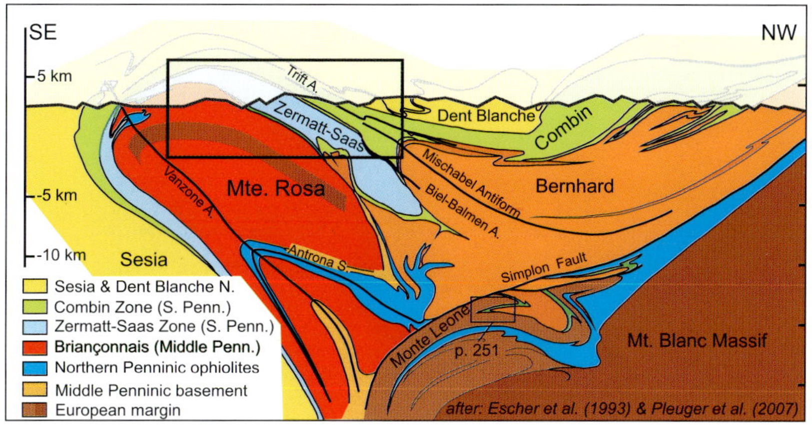

Geological cross section through the Mte. Rosa region. The black box shows the region of the photo at left. Parts of the cross section that have long been eroded are shown transparently.

The Monte Rosa east face reaches from 2200 m above sea level at its base to 4500 m elevation at the summit of Dufourspitze (4634 m), the highest peak of the Mte. Rosa Massif. The face is considered to be the highest face in the Alps. Geologically, the face is part of the Monte Rosa Massif – one of the three classic *Internal Massifs* of the Alps. The massif is made up of granites and gneisses that formed in the European crust between 260 and 310 million years ago (Variscan times). Alpine tectonics disconnected the massif from its base at about 40 Ma. It was then dragged to a depth of about 50 kilometres where it was heated to roughly 500° C. Under these conditions, minerals like feldspar, biotite and white mica were metamorphically changed to garnet, phengite and albite. Today the massif lies surrounded by rocks of the southern Penninic Ocean floor (Zermatt-Saas Zone) in the centre of the Penninic domain of the Alps.

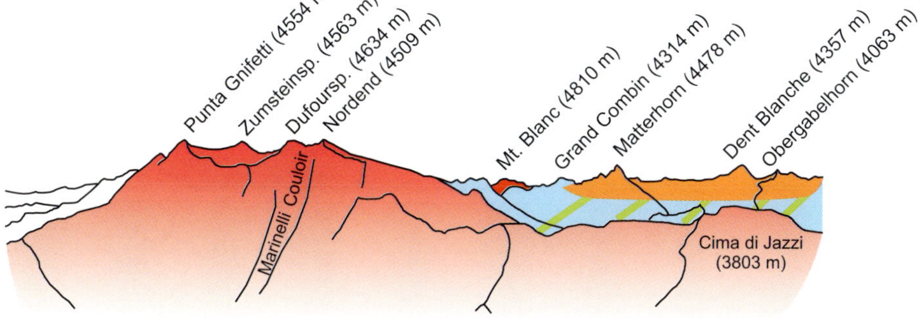

Sketch of the photo at left. The colour code corresponds to the geological cross section above.

2.3. Domain of the Penninic Oceans

Background on p. 67: The slopes of Mont Emilius (3 559 m) on the south side of the Aosta Valley. The valley crosses one of the largest interconnected regions of rocks in the Alps that once were part of the Penninic Ocean floor. Near the town of Aosta, these rocks include predominantly calcareous schists that once were sediments on the floor of the southern branch of this ocean: the Piemont-Ligurian Ocean. However, at Mont Emilius itself, continent-derived Austro-Alpine rocks lies on top of these ocean-derived rocks. This klippe is very similar to the Dent Blanche Nappe.

Map on p. 67: Tectonic map of the Western and Central Alps showing the major geological units of the Penninic Ocean domain of the Alps. Valais Ocean = Northern Penninic; Briançonnais = Middle Penninic; Piemont-Ligurian Ocean = Southern Penninic. The map corresponds to the one shown on p. 55 where all units of the Penninic domain are collectively shown in blue.

Right: The Misox Valley in the Swiss canton Ticino. The rocks in the Misox Valley are metamorphosed sediments and volcanics that once were part of the northern Penninic Ocean basin. The remnants of this ocean floor here form the so-called Misox Zone. Structurally, this zone lies above the Adula Nappe (on the west side of the valley, in the picture at left) and below the Tambo Nappe (in the photo at right). The former belongs to the Sub-Penninic units derived from the European margin, the latter is made up of rocks of the Briançonnais Fragment.

The Mesozoic *Penninic Ocean* is also called the *Alpine Tethys*. The ocean is named after the *Pennine Alps* (or: *Valais Alps*) because the mountains there are largely built up of rocks from the floor of this ocean. The ocean was created in connection with the first formation of the Atlantic Ocean in the Jurassic. At this time, the Atlantic Ocean propagated successively northwards and tore South America and Africa apart. When the tear reached the latitude of Gibraltar, it changed its direction towards the northeast to form the Penninic Ocean. Only once this ocean was subducted beneath the Adriatic plate, tens of millions of years later, did the Atlantic continue to propagate to the north to separate Europe from North America.

When talking about the *Penninic Oceans*, one should usually do so in plural, as there were several oceanic basins separated by fragments of rocks that identify them to be of continental origin. Geologists interpret these fragments as islands and peninsulas between the ocean basins. The *Briançonnais Fragment* is the largest of these fragments. Its rocks constitute the Suretta-, Tambo- and St. Bernhard Massifs. The rocks of the Penninic ocean floor itself are particularly wide-spread in the Central Alps: *ophiolites*, *gabbros*, *basalt* and other dark quartz-free rocks of magmatic origin. Many of these rocks were metamorphosed to amphibolites during Alpine tectonics, but their bulk rock chemistry still identifies them clearly as part of a former ocean floor. However, much more common than the magmatic rocks of the oceanic crust itself are the sedimentary rocks that were deposited on the deep sea floor. In particular these are calcareous muds that were later metamorphosed into the well known *Bündnerschiefer.* Other deep water sediments occur in between.

Europe
Basement (Black Forest, Vosges)
External massifs, Sub-Pennin. nappes, (Ultra)Helvetic nappes, Dauphinois
Valais Ocean
Lower (Northern) Pennin. nappes
Rhenodanubian Flysch
Briançonnais Fragment
Middle Penninic Nappes
Piemont-Ligurian Ocean
Upper (Southern) Penninic oceanic rocks
Cervinia Terranes
Upper Pennin. continental rocks
Adriatic Plate
Austro-Alpine nappes
South Alpine
Periadriatic granitoids
Apennin
Sedimentary cover of the Foreland, Po Basin
100 km
Vosges
Black Forest
Rhine
Basel
Zurich
Molasse Basin
Wägital
Falknis
Sulzfluh
Jura
Bern
Schlieren
Mythen
Gurnigel
Gotthard
Arosa
Tasna
Engadine Window
Niesen
Adula
Platta
Leponine Dome
Margna
Bellinzona
Geneva
St. Bernard
Mt. Blanc
Dent Blanche N.
Matter-horn
Mte Rosa
Misox Zone
Bergell
Adamello
Rhône
Sesia Zone
Gran Paradiso
Biella
Milan
Verona
Grenoble
Lanzo Peridotites
Po Basin
Turin
Monferrato
Po
Dora-Maira
Voltri Ophiolites
Sestri-Voltaggio
Ligurian Apennin
Genova
Mediterranean Sea
Nice
after: Froitzheim et al. (2008)

Right: The Lukmanier Pass forms the boundary between the Swiss cantons of Ticino and Grisons and is the lowest crossing of the principal drainage divide of the Swiss Alps. In Roman times, it was an important trade route. Geologically, the Lukmanier Pass is at the contact between the Bündnerschiefer of the North Penninic Ocean (Valais), the gneisses of the Gotthard Massif and the high grade metamorphic rocks of the Lepontine Dome.

Page 69: At the end of the Blenio Valley in Ticino the front of the Sub-Penninic Adula Nappe lies on top of the enormous masses of calcareous sediments (Bündnerschiefer) that once were deposited on the floor of the Mesozoic North Penninic Ocean. The triangular peak in the left third of the photo is called Sosto (2 220 m). It is completely made up of the monotonous Bündnerschiefer. On the skyline the granites and gneisses of the Gotthard Massif are visible.

Below: North-south cross section through the Adula Nappe, some kilometres east of Lukmanier Pass. It may be seen how the nose of the nappe pushes into the Bündnerschiefer of the Valais Ocean. The red ring shows the position of the photo on p. 69.

A. North Penninic Ocean (Valais trough)

The North Penninic Ocean (also called: Valais Ocean) was a west-east oriented ocean basin that formed about 90 – 100 million years ago as the northernmost and last basin of the Mesozoic system of Penninic oceans. Because of its geographic position it was also the last basin that disappeared much later (in the Neogene) underneath the Adriatic plate. As a consequence, the rocks of its ocean floor ended up as one of the lowest units in the Alpine pile and its therefore also often called Lower Penninic.

Two important rock types belong to this ocean: (i) the rocks from the Penninic ocean floor itself and (ii) the somewhat younger sediments that were deposited into the deep trench that had formed during the subduction of this ocean.

Typical example of the rocks from the Penninic ocean floor are the *Bündnerschiefer*: calcareous deep water sediments of a variable degree of metamorphism. They are common rocks in the namesake Swiss canton Grisons and occur often mixed with the rocks of the oceanic crust itself (ophiolites). They are also found in the *Engadine Window* in Tyrol in Austria and in the *Glockner Nappe* of the *Tauern Window*. A narrow stripe of these rocks occurs along the Misox Valley and surrounds the Lepontine Dome (p. 67). From the upper end of the Leventina Valley, these rocks lead to the Lukmanier Pass and then westwards across the Nufen Pass, through the Blinnenhorn into the Rhône Valley, pass the south side of Mt. Blanc and then all the way to the eastern end of the Pelvoux Massif.

A typical example of the later trench deposits is *Flysch* a deep see sediment mostly made of marls with signs for active deposition. The *Rhenodanubian Flysch* forms a narrow band of rocks along the entire northern margin of the Eastern Alps through the Forggensee, Salzburg, Attersee and Traunsee to north of Vienna. The Rhenodanubian Flysch has barely been metamorphosed and is currently being overthrust by the Northern Calcareous Alps.

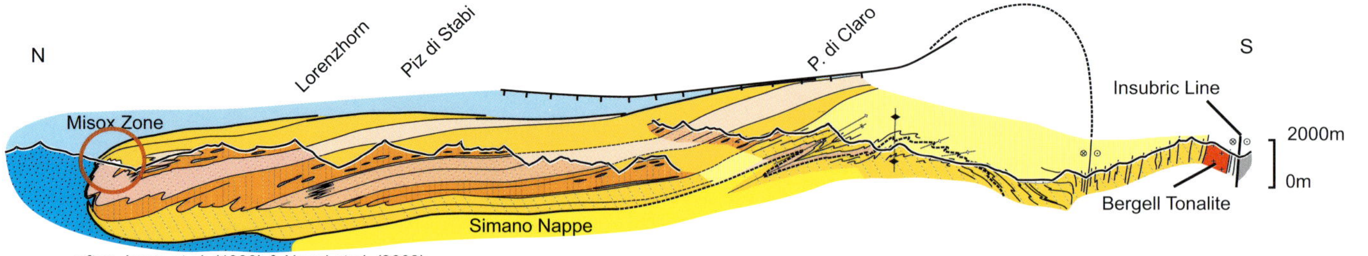

after: Jenny et al. (1923) & Nagel et al. (2002)

Front of the Adula Nappe
Gotthard Massif
Bündnerschiefer
Adula Nappe
Recent sediments

B. Middle Penninic Continent

The Middle Penninic continent is also called the *Briançonnais Fragment.* Its rocks have a type locality around the city of Briançon in the French Alps. The rocks are granites, metamorphosed river sediments and shallow marine sediments that are all definitely of continental origin. However, the rocks are always over- and underlain by rocks from the Penninic Ocean, so that the package of rocks is interpreted as an island in the system of Penninic ocean basins. Rocks of this *Briançonnais Fragment* occur in large parts of the Central and Western Alps and many of them are sheared off their base and transported long distances to the north. The limestones of the Rätikon region near the Swiss-Austrian border, the *klippen* of the Mythen or the Préalpes north of Gstaad, or the Suretta and Tambo nappes and the St. Bernhard Nappe of the Central

Alps (Mischabel Group) are part of the *Briançonnais Fragment*. From the Mischabel Group a broad region of Middle Penninic rocks leads westwards between Mt. Blanc and Gran Paradiso and reaches the Italian coast near Savona.

Above and right: The Mischabel Group in the Central Alps is made up of rocks of the Briançonnais Fragment (orange on sketch of photo at right). North of the Alphubel these rocks lie atop of the southern Penninic Zermatt-Saas Zone (green). Monte Rosa in the background at left, however, is part of the Sub-Penninic gneisses, and the peak of the Matterhorn on the skyline is part of the Dent Blanche Nappe.

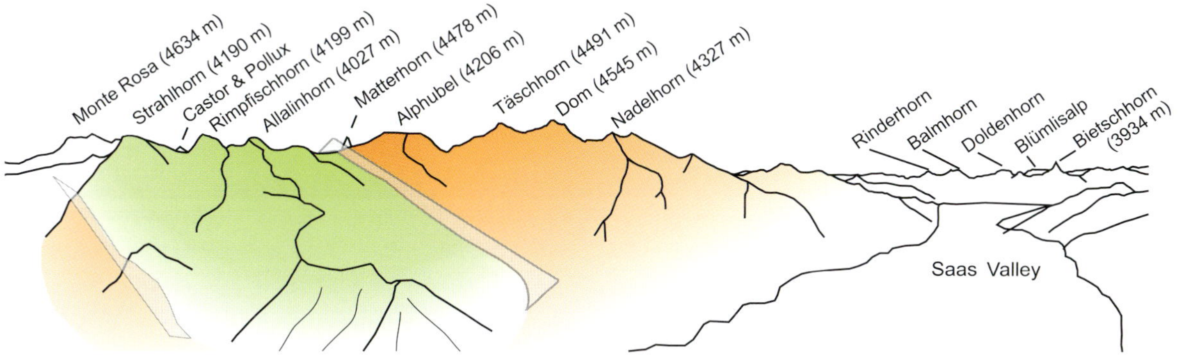

Above: Grand Combin seen from the west. The large valley in the foreground is the Val d'Entremont. The highway to the St. Bernhard Pass leads along it. Grand Combin is built up to a large extent of oceanic rocks derived from the South Penninic Ocean. On the side seen here, it is made up of basement of the Mont Fort Nappe.

Right: Geological cross section through the Central Alps showing that the rocks of the former Piemont-Ligurian Ocean completely envelop the Monte Rosa Massif (legend on p. 258).

C. South Penninic Ocean (Piemont)

The southern Penninic ocean is also called the *Piemont-Ligurian Ocean*. It initially formed in the Jurassic (140 – 165 Ma) and disappeared again around 40 – 50 Ma ago underneath the Adriatic plate. Because of its palaeogeographic position as the southernmost branch of the system of Penninic oceans (p. 48) it remained high in the tectonic stack of the Alps during Alpine deformation. Its rocks are therefore referred to as the Upper Penninic.

Rocks of the Upper Penninic build up most of the Alps southwest of the Valais Mountains, east of Aosta and east of Briançon. Only the Sub-Penninic tectonic windows (Gran Paradiso, Dora Maira) are excluded. As for the rocks of the Valais Ocean, the rocks of the Piemont-Ligurian Ocean are deep sea

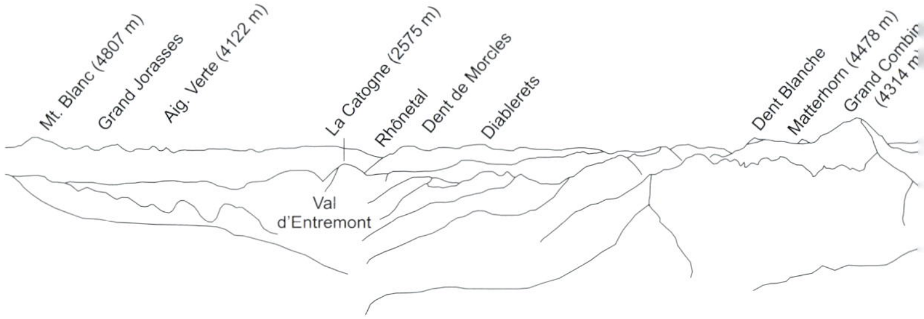

sediments (Bündnerschiefer), gabbros, basalts and ophiolites of the former oceanic crust. In between there are occasional splinters of continent derived rocks, for example in the Sesia Zone, in the Dent Blanche Nappe or the Margna Nappe. These rocks are interpreted as islands in the South Penninic Ocean: the *Cervinia terrane (Margna unit)*. During later subduction of the Penninic ocean floor and the subsequent Alpine defor-

mation, the rocks of the Piemont-Ligurian ocean floor were broken up into several tectonic nappes that were then thrust over the top of each other.

The *Zermatt-Saas Zone* is one of the lowermost of these nappes. Consequently, it experienced a relatively high degree of metamorphism. The zone consists of serpentinites and eclogites and other ophiolitic rocks that build up the Allalinhorn in the Mischabel Group, the rocks around Zermatt, the base of the Matterhorn and the Breithorn. The rocks surround the Gran Paradiso Massif and lead southwards into the Italian Alps. Above the Zermatt-Saas Zone (and below the Dent Blanche Nappe) lies the *Combin Zone* which builds up Grand Combin and leads on (always just east of the Briançonnais rocks) all the way to the south until it disappears underneath the Po basin near Cuneo.

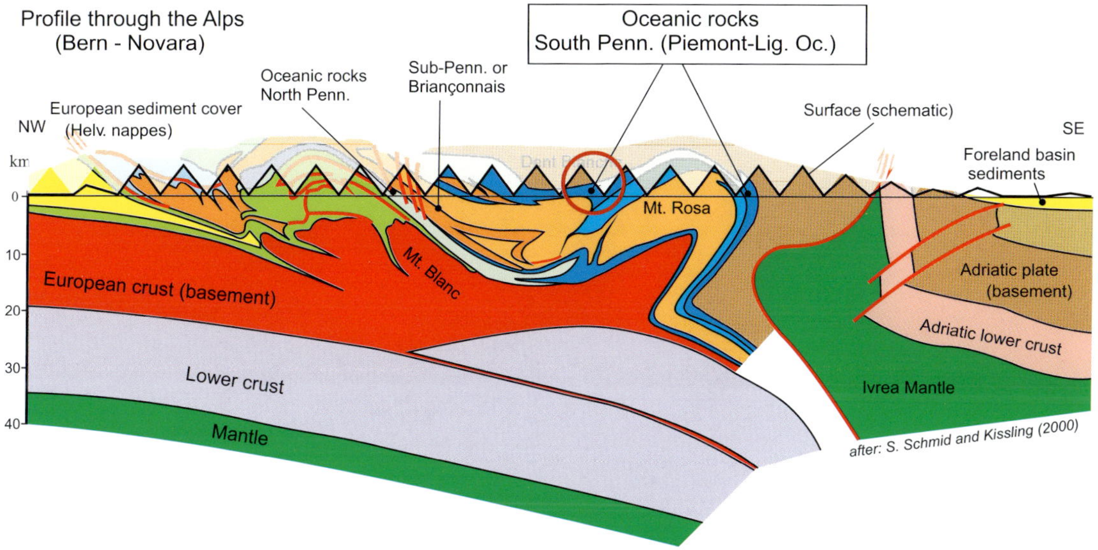

D. Penninic Rocks of the Eastern Alps

In the Eastern Alps, rocks derived from the Penninic Ocean floor may only be seen in three tectonic windows through the Austro-Alpine rocks and in the *Rhenodanubian Flysch Zone*. The tectonic windows include the *Engadine Window*, the *Tauern Window* and the *Rechnitz Window*. In these three places the Austro-Alpine rocks were tectonically pulled aside, like curtains from a window, so that the rocks from the underlying former Penninic Ocean floor have become visible. The rocks are mostly monotonous, calcium-rich, deep sea sediments that are known in the Central and Western Alps by the name of "Bündnerschiefer" or *"Schistes Lustres"*. The rocks are of medium metamorphic grade. Chunks of dark, quartz-poor rocks of the former oceanic crust lie between. In the Tauern Window this mix of sediments and fragments of oceanic crust are called Glockner Nappe (p. 158). In the Engadine Window these rocks may be subdivided into Lower Penninic, Middle Penninic and Upper Penninic nappes. In the other tectonic windows this subdivision cannot be made. The Tauern Window is the only one of the three Penninic windows in which erosion has cut deep enough so that the European crustal basement underneath has been exposed: the *Zentralgneise*. The *Rhenodanubian Flysch Zone* is a long stripe of Penninic rocks that parallels the entire northern margin of the Eastern Alps between the Northern Calcareous Alps and the Molasse Zone. The rocks are a mixture of marls, mudstones and sandstones and were deposited in water depths between 3000 and 5000 m. They were deposited into a deep sea trench that formed in connection with the subduction of the Penninic Ocean.

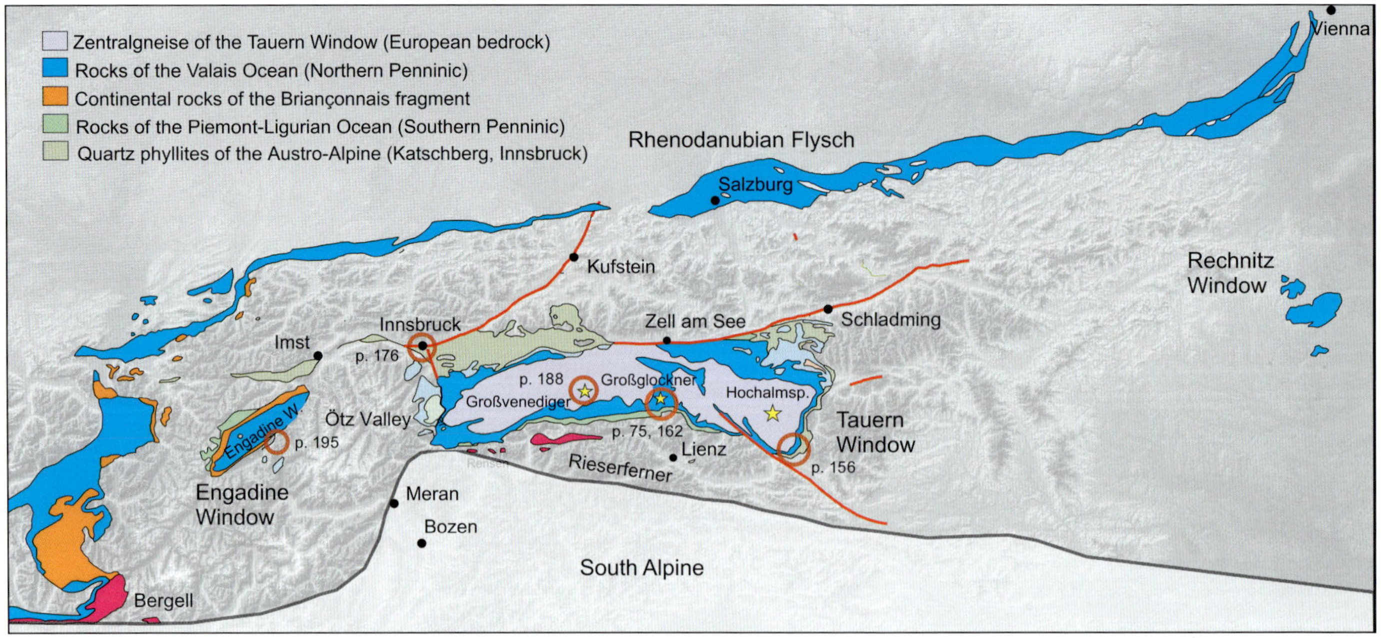

Right: Map showing the rocks of the Eastern Alps that once belonged to the Penninic Ocean in the Mesozoic. Periadriatic intrusive rocks (red) are shown for orientation.

Page 75: Großglockner (3798 m) is the highest peak of Austria. Ironically, it is made up of rocks that are lower origin than most other rocks of the Eastern Alps: they stem from the floor of the Mesozoic Penninic Ocean. The photo is looking west from a point high above the Großglockner Hochalpenstraße. In the foreground is one of the green serpentinite bodies that often lie between the Bündnerschiefer of the Penninic Glockner Nappe. The triangular peak in the middleground is the Fuscherkarkopf (3331 m). The layering of the quartz-rich phyllites of the Bündnerschiefer may be well seen there.

2.4. Adriatic Plate

The photo at right shows the Rottenmanner Tauern with the Bösenstein (2448 m) in the centre. The rocks here are mica schists and amphibolites as typical for the Austro-Alpine basement. On the sky line the Seetaler Alps with their highest summit, the Zirbitzkogel (2396 m).

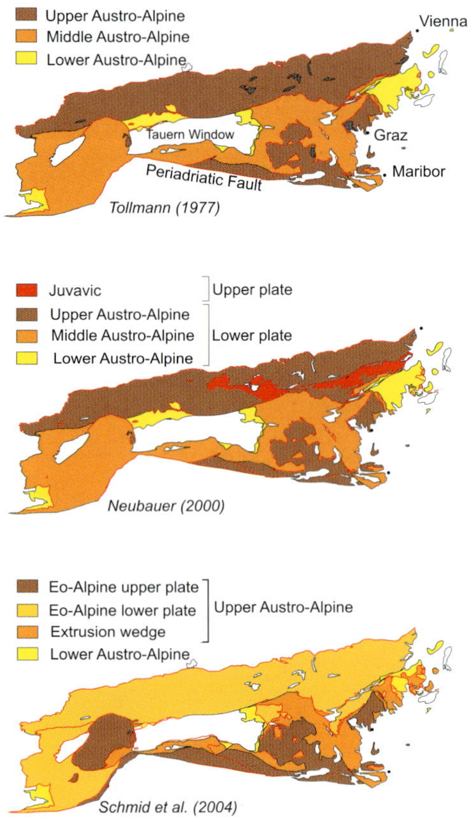

Subdivision of the Austro-Alpine according to different authors. The subdivision according to Schmid et al. (2004) is generally followed in this book, although the Tollmann subdivision is explained on this page.

A. Austro-Alpine

The Austro-Alpine is the tectonic name for most rocks of the Alps that lie north of the Periadriatic Line and east of a north-south line from the Rätikon region through Arosa to the Bergell Massif. Only the rocks of the Tauern-, Engadine- and Rechnitz Windows and the Flysch Zone are excluded from this. Palaeogeographically, the Austro-Alpine rocks are derived from the Adriatic (Apulian) plate that once lay in the Mesozoic between the Penninic Ocean to the north and the Meliata-Hallstatt Ocean to the south.

The Austro-Alpine includes a large variety of rocks, many of which were repeatedly metamorphosed during the Variscan, Permian and Alpine tectonic processes. Because of this, large parts of the Austro-Alpine are high grade metamorphic rocks. In a somewhat older but classic subdivision, the Austro-Alpine is subdivided into three parts:

The Lower Austro-Alpine is the lowest tectonic unit. It is only visible where the Austro-Alpine lies directly on the Penninic units of the Alps, for example along the western margin of the Tauern Window. On the Wechsel Pass south of Vienna it forms an envelope around the *Rechnitz Window.* The largest exposures of the Lower Austro-Alpine occur along the westen margin of the Eastern Alps, where the Austro-Alpine lies on the Penninic of the Central Alps: in the *Bernina* and *Err Nappes*.

The Middle Austro-Alpine includes the bulk of the metamorphic rocks of the Eastern Alps: the Ötztal Alps, Schladminger Tauern, Glein- and Koralpe, Kreuzeck Massif and many others. The typical rocks of the Middle Austro-Alpine are garnet mica schists, amphibolites, marbles and various gneiss types. The Middle Austro-Alpine is usually high grade metamorphic.

The Upper Austro-Alpine constitutes the tectonically highest parts of the Austro-Alpine. This includes the Northern Calcareous Alps, the Paleozoic of Graz and the Greywacke Zone. Because these rocks were the tectonically highest units, the rocks were barely metamorphosed and limestones are often fossil bearing.

In the last years, geologists recognised that the Alps are better explained if the Middle- and Upper Austro-Alpine are not separated, because their present day relative position does not necessarily reflect their respective belonging to the upper or lower plate during the Cretaceous tectonics.

Right: Typical mica schist of the Austro-Alpine basement (Anger Basement, Styria, Austria). The crystals are calcium-rich amphiboles (Hornblende). They formed around 80–90 Ma at about 600°C and in 30–40 kilometres depth during the Eo-Alpine metamorphism of the Upper Austro-Alpine basement (formerly: Middle Austro-Alpine) (photo: R. Schuster).

1 EURO

Hochgolling (2 862 m) is the highest peak of the *Lower Tauern* and is typical for the variegated nature of the Austro-Alpine basement. The mountain itself is made up of a colourful mixture of high grade metamorphic mica schists, amphibolites and plagioclase gneisses. These rocks belong tectonically to the *Silvretta-Seckau Nappe system* of the Upper Austro-Alpine nappes. This basement suite also includes much of the Schladminger Tauern and the Gleinalpe. The *Lungauer Kalkspitzen* (limestone peaks) can be seen behind the Hochgolling. In contrast to the Northern Calcareous Alps, they are in direct stratigraphic contact with the crystalline basement underneath and are therefore similar to the Ortler Massif (p. 200). They are considered to belong to the Lower Austro-Alpine nappes. The Dachstein Massif in the background is part of the Northern Calcareous Alps and thus part of the Upper Austro-Alpine nappes. The rounded cirques (e.g. the Klafferkessel in the photo at right) were carved during the glaciation periods. In total, erosion has removed several tens of kilometres of rocks from the tops of these mountains during the last 30 Ma.

Profile through the Lower Tauern (Koralpe-Wölz Complex, Upper Austro-Alpine crystalline)

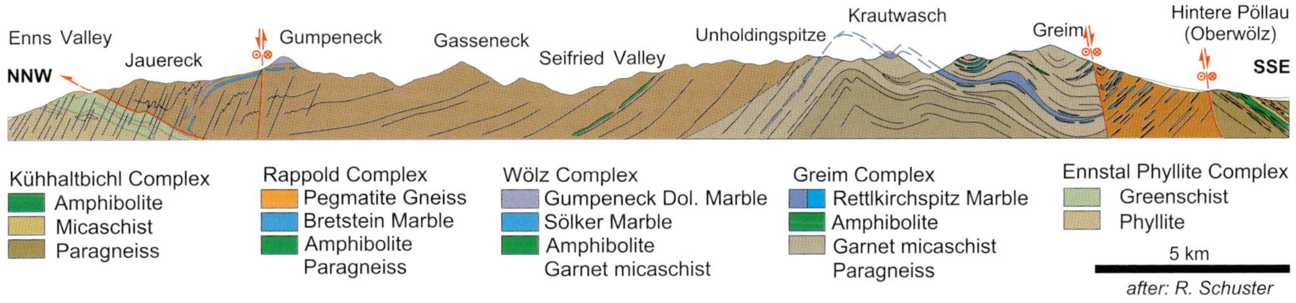

B. South Alpine Zone

The South Alpine zone includes all rocks of the Alps that lie south of the Periadriatic Line as the separating suture between South Alpine- and Austro-Alpine Zones (p. 110). This suture zone follows a series of distinct valleys from west to east along the entire chain of the Alps. However, palaeogeographically, the South Alpine was once connected to the Austro-Alpine, with both being part of the Mesozoic Adriatic plate. Thus, the South Alpine and Austro-Alpine contain very similar rocks: high grade metamorphic rocks, Permian sediments, volcanics and massive carbonate platforms from the Triassic and Jurassic. The South Alpine and Austro-Alpine were only separated around 35 Ma when the Periadriatic Line formed. The most important mountain ranges built up by South Alpine rocks are the *Bergamo Alps*, the *Adamello Massif*, the *Brenta Group*, *Dolomites*, *Julian Alps*, *Carnic Alps*, the South *Karavanke* and the region around the large north Italian lakes.

One major difference between the South Alpine and the Austro-Alpine zones is that the former was less involved in the Alpine tectonic processes and is therefore barely overprinted metamorphically. As a consequence, large parts of the South Alpine preserve the Permian and Variscan evolution of Europe. The 200 Ma old flat-lying carbonates of the Dolomites document this beautifully: they illustrate that little pervasive deformation has occurred in the last 150 Ma in the region. Nevertheless, since the formation of the Periadriatic Line around 35 Ma, the South Alpine has been thrust slowly southwards over the top of the Po basin along large flat-lying thrust zones like the *Valsugana* or *Wontello Thrust*.

The western part of the South Alpine is somewhat more deformed than the eastern part. It is called the *Lombardic-Guidicaric fold belt*. This fold belt includes in the very west the *Canavese Zone*: a narrow band of crystalline rocks between branches of the Periadriatic Line. The *Ivrea Zone*, the *Strona-Ceneri Zone* and the *Val Colla Zone* are the next successive zones towards the southeast. Between Lago di Como and Lago di Garda lies the *Orobic Basement*. The Orobic Basement is of Variscan age, but was folded in Alpine times into large open folds. To the south, it disappears underneath Mesozoic carbonates.

East of Lago di Garda the South Alpine Zone is barely deformed. Just as it was 100 million years ago, there is an undisturbed sequence with Variscan gneisses at the base, thick Permian volcanic sequences above and Triassic, Jurassic and Cretaceous carbonates of the Dolomites and Julian Alps at the top of the pile. The Carnic Alps and the southern part of the Karavanke range are also made up of carbonates, but are significantly older (p. 94).

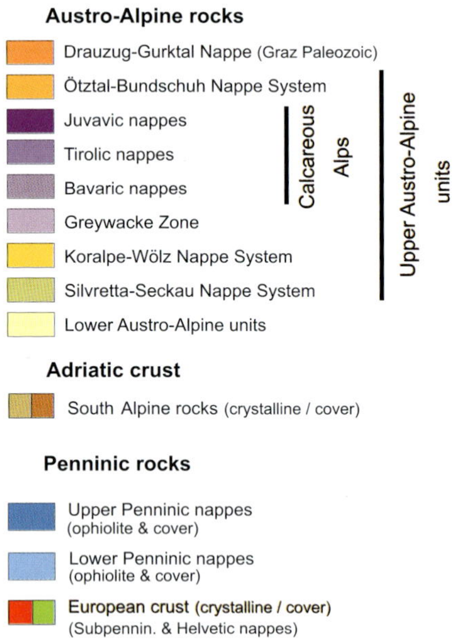

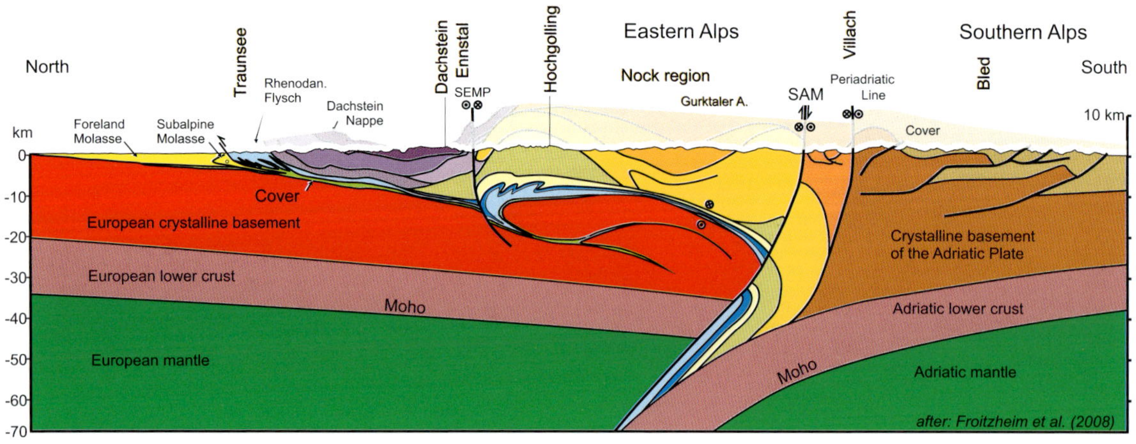

Geological north-south cross section through the Eastern and Southern Alps approximately from the Traunsee to the Woheiner Lake in Slovenia. The white line is the present day topography; eroded parts of the profile are shown transparently.

Marmolata
Sella
Civetta
Val Cismon
Sass Moar (2814 m)
La Fradusta (2993 m)
Pala di San Martino (2982 m)
Val Pradidali

Above: The Bergamo Alps. The narrow road in the foreground leads across the San Marco Pass, from Pizzatorre to the valley of the river Adda (Valtelino) near the northern end of Lago di Como (middle ground). In the left background the Po basin may be seen. In the centre of the photo the Monte Rosa Massif can be seen on the skyline. To the right the mountains of the Bernese Oberland. At the very right of the photo, the light coloured ragged peaks are the Bergell Massif with the dark peak being Monte Disgrazia.

The Orobic Basement of the Bergamo Alps is typical for much of the crystalline basement of the South Alpine Zone. It is made up of high grade metamorphic gneisses that were heated and metamorphosed during the Variscan mountain building processes. In the Permian they were buried by volcanic rocks. During Alpine orogenesis there was only weak metamorphic overprinting and the cover rocks were eroded. The middle crust was exhumed to the surface by overthrusting along major south-directed thrust zones. South of the Orobic Basement these Mesozoic cover rocks are still preserved.

Right: Geological map of the Southern Alps. The series of fault zones that limit the Southern Alps to the north are collectively known as the Periadriatic Line.

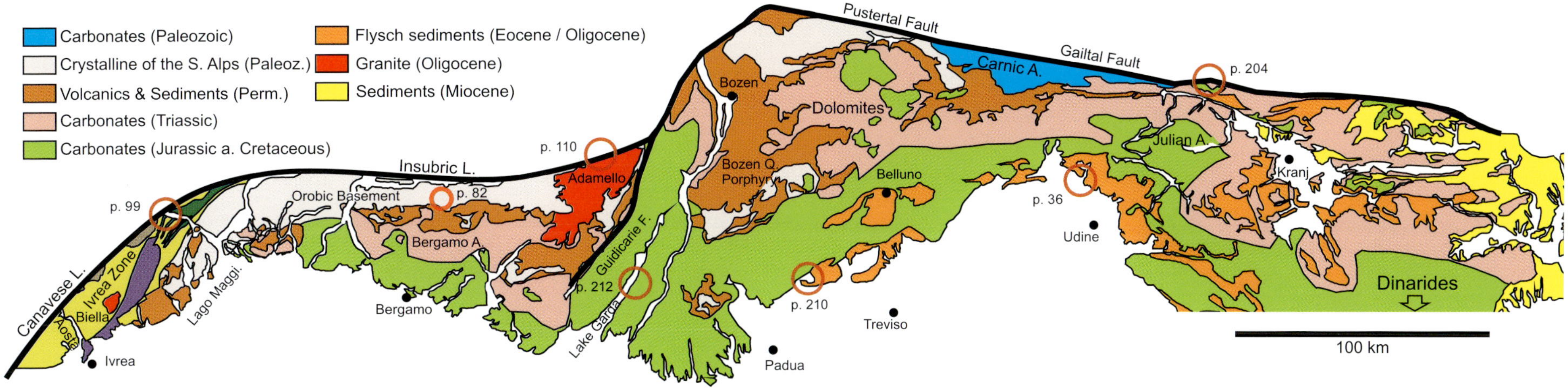

Carbonates (Paleozoic)
Crystalline of the S. Alps (Paleoz.)
Volcanics & Sediments (Perm.)
Carbonates (Triassic)
Carbonates (Jurassic a. Cretaceous)
Flysch sediments (Eocene / Oligocene)
Granite (Oligocene)
Sediments (Miocene)
Pustertal Fault
Gailtal Fault
Carnic A.
p. 204
Bozen
Dolomites
Julian A.
Kranj
p. 110
Adamello
Insubric L.
Orobic Basement
p. 82
Bozen Q.
Porphyry
Belluno
p. 36
Udine
p. 99
Bergamo A.
Guidicarie F.
Canavese L.
Ivrea Zone
Biella
Lago Maggi.
Bergamo
p. 212
Lake Garda
p. 210
Treviso
Padua
Ivrea
Dinarides
100 km

3. Tectonic Evolution

Age of metamorphism:

high-temperature metamorphism
- 0-59 Ma (<300°C)
- 0-59 Ma (>300°C)
- 60-110 Ma (<300°C)
- 60-110 Ma (>300°C)
- 0-59 Ma cooling ages overprinting 60-110 Ma old metamorphism

high-pressure metamorphism
- 35 - 59 Ma
- 60 - 89 Ma
- 90 - 110 Ma

Rocks:
- Granitoids (25-41 Ma)
- unmetamorphosed pre-Mesozoic rocks
- unmetamorphosed post-Mesozoic rocks

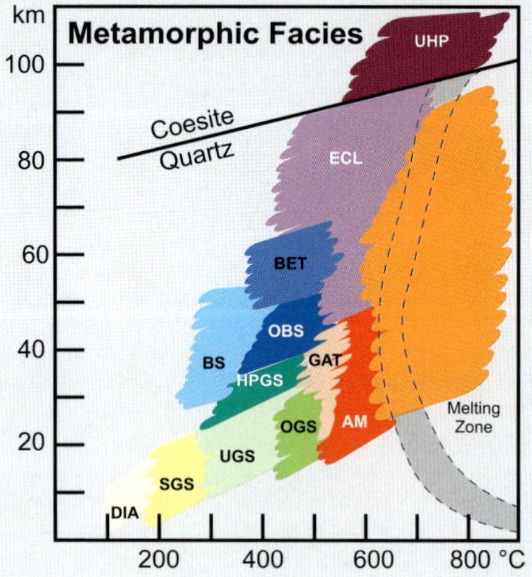

Grade of metamorphism:

- Diagenesis (DIA)
- Sub-greenschist facies (SGS)
- Lower-greenschist facies (LGS)
- Upper-greenschist facies (UGS)
- HP-greenschist facies (HPGS)
- Greenschist - Amph. transition
- Amphibolite facies (AM)
- Blueschist facies (BS)
- Upper-blueschist facies (UBS)
- Blueschist-eclogite transition
- Eclogite facies (ECL)
- Ultra-high pressure facies (UHP)
- various high-*T* facies (purple stars = eclogite occurrence)
- Amphibolite f. overprinting
- Tertiary granites

Metamorphic Grade of the Rocks of the Alps

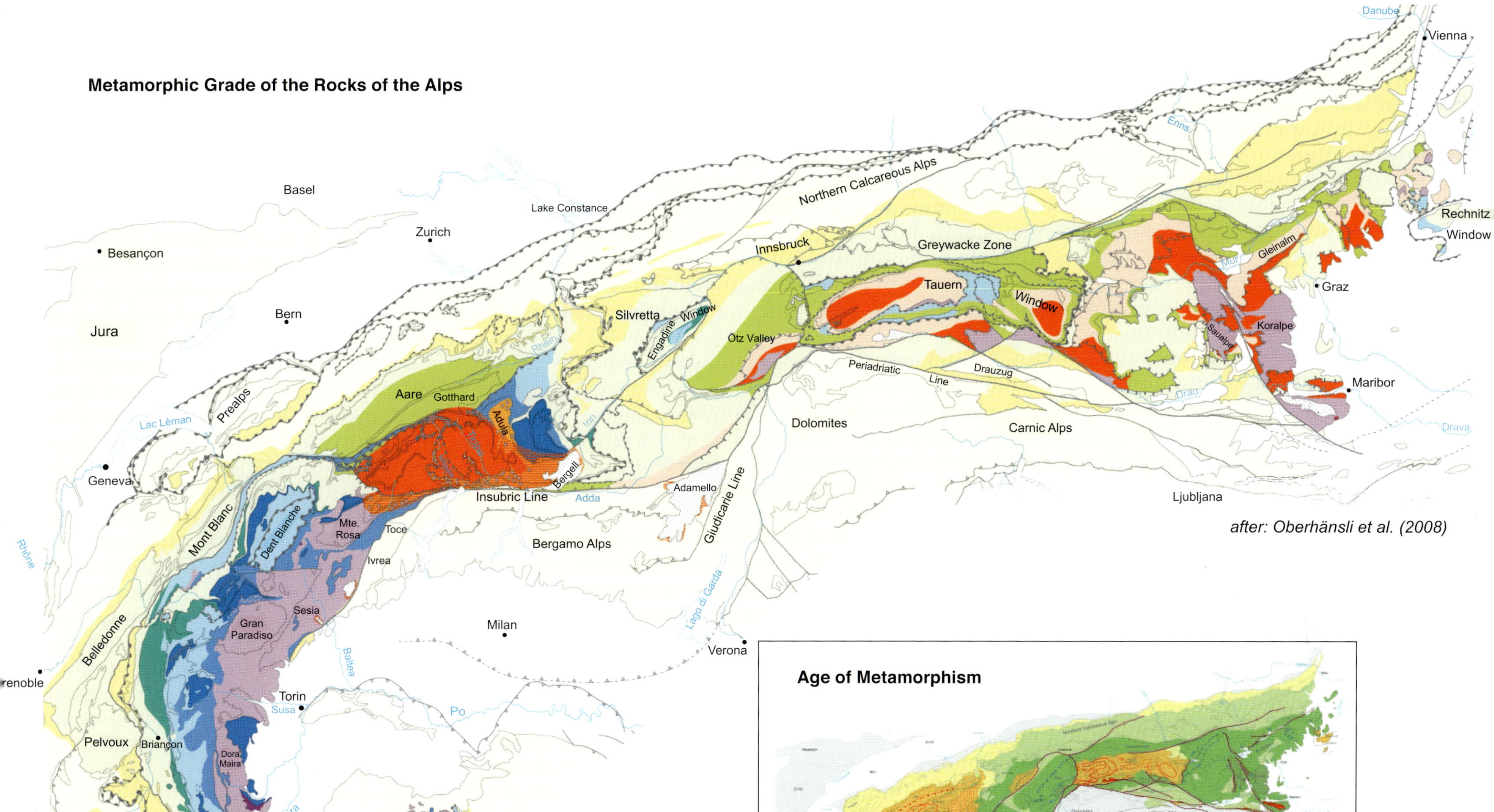

after: Oberhänsli et al. (2008)

Age of Metamorphism

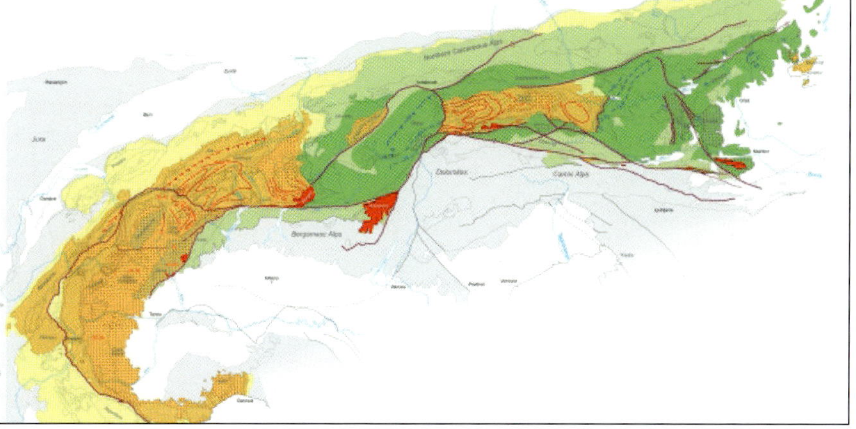

Overview

The chapter tells the story of the tectonic evolution of the Alps. The first few pages give an overview starting with the oldest rocks of the Alps and finishing with the present day topographic development. From p. 94 onwards, one or two double pages are dedicated to each geological period.

The oldest rocks in the Alps are the carbonates and schists of the Carnic Alps, the Grazer Bergland, and few other small regions in the Eastern Alps. These rocks were once deposited as sediments along the margin of Gondwana some 400 – 500 million years ago (p. 10). Little is known about the shape or origin of these ancient continental margins and the Alps existed neither as a region nor as a mountain range at this time. All parts of today's central Europe were still widely distributed around the entire globe.

The Variscan (often also called Hercynian) mountain building phase (orogeny) affected large parts of Europe, the Atlas mountains and the Appalachians around 380 – 340 million years ago. The rocks in the Black Forest, the Bohemian Massif, the French Massiv Central and the Bretagne were formed. Around 310 Ma, this active tectonic phase was over and the Variscan mountain range eroded. The landscape slowly flattened. In the Alps, only the roots of this Variscan mountain belt remain in place left untouched by much later Alpine events, for example in the Wachau or the Waldviertel regions of Austria. Other than that, there are many granites and gneisses that formed during the Variscan orogeny but were later intensely kneaded, folded and re-heated during the Alpine events. Although very much changed, these rocks often contain minerals or structures that are more than 300 million years old. This includes the gneisses that build up the highest peaks of the Hohe Tauern range, as well as parts of the Austro-Alpine and the gneissic cores of the Aare, Mt. Blanc, Pelvoux or Argentera Massifs.

In the *Permian* (about 290 Ma) a series of new rocks started forming in the Alps in connection with a wide spread north-south directed stretching of the continent. The lithosphere was extensionally thinned and melts formed in the Earth's uppermost mantle. Dark, silica-poor melts intruded the lower crust and crystallised as gabbros. Today these gabbros are found in the Kor- and Saualpe, in the valleys east of Bergell or in the Ivrea Zone near Lago Maggiore. The rising magmas caused partial melting in their surroundings forming granites and pegmatites that intruded somewhat higher into the middle crust and caused metamorphism of the surrounding rocks at up to 650°C. Pegmatites of this age can be found in the Kreuzeck Group or the Koralpe, where they host Europe's largest lithium deposit. Some of this magma even reached the surface and caused widespread volcanism. The Bozen quartz porphyry (a popular building stone), the rhyolites of the Glarus Nappe in the Swiss Helvetic region (Verrucano), or the volcanics of the St. Bernhard Nappe are all part of this suite of volcanic rocks that formed in the Permian.

The ongoing stretching of the lithosphere flattened the surface of the crustal fragments that would become the Alps much later. Flat regions with gigantic river deltas formed. The rivers drained a landscape with a dry climate and poured their sedimentary load in long east-west oriented valleys. These basins formed along fractures that formed in response to the north-south stretching of the continent. The quartzites in the Semmering region, some of the mica schists of the Seckauer- and Schladminger Tauern and the thin Permian sediment mantle on top of the gneisses of the Aare Massif are examples of rocks that were deposited as sandy and silty sediments in the Per-

Below and far right: Earth in the Carboniferous and Permian. The convergence of the southern continent with Laurussia caused closure of the Palaeotethys ocean and finally the Variscan mountain building phase (red region below). In the Permian roughly the same region (red region on globe at far right) was re-stretched during the opening of the Meliata-Hallstatt-Ocean (MHO) (after: McCann 2008).

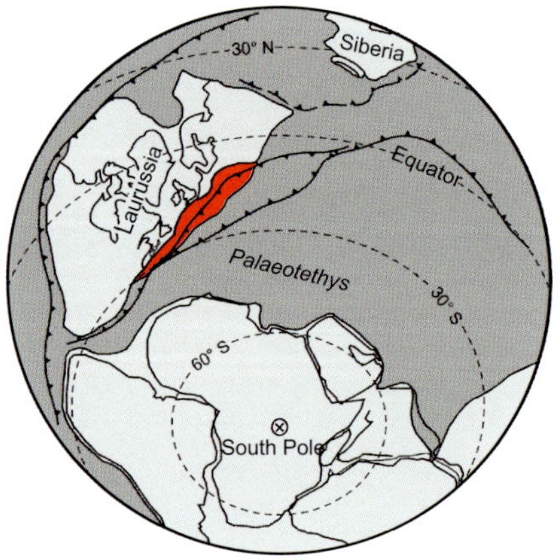

Early Carboniferous (ca. 340 Ma)

mian. They were transformed into metamorphic rocks only much later.

About 250 million years ago this north-south directed stretching suddenly stopped in the entire Alpine realm. A new ocean progressed from the east into the region south of the Alps: the *Meliata-Hallstatt Ocean*. This ocean is the westernmost embayment of the Tethys Ocean that covered half of the globe at the time. This ocean kept extending westwards, and when it finally flooded the region south of central Europe the continent to its north became tectonically quiet.

A long phase of static cooling and sagging of the continent began. Eventually, the surface of the continent reached sea level and large lagoons and brackish lakes formed. The salt that was deposited in these lagoons today forms all the major salt deposits of the Alps, for example those in the Salzkammergut region. In the Dolomites, the fossil-bearing Bellerophon beds were deposited and somewhat later the Werfen Formation of the Northern Calcareous Alps. Both are distinctive rock types that mark the beginning of the reef development that characterised the following Triassic period.

During the Triassic, the land sank further and further down and completely drowned: the salty lagoons became a coral sea on the continental shelf along the northern margin of the Meliata-Hallstatt Ocean. Because the sinking did not stop, corals kept growing towards the light and eventually formed kilometre thick reefs that build up the *Dolomites* and the *Northern Calcareous Alps* today. This evolution lasted for almost 50 million years until it was interrupted in the Jurassic by new tectonic events. In the Jurassic, the *Atlantic Ocean* began to open in the west.

About 165 million years ago a branch of this new ocean began to rip eastwards forming a series of valleys and grabens north of the Alpine region. Soon an ocean basin formed there: the *Penninic Ocean* was born (also called *Alpine Tethys*). This ocean now separated Europe from Africa, and both continents began to drift in opposite directions. The rocks of this Penninic ocean floor form much of the Central and Western Alps west of the Rhine Valley in Switzerland. Between the new ocean to the north (Penninic Ocean) and the Permian ocean to the southeast (Meliata-Hallstatt) the African plate now had an appendage, the so called *Adriatic spur*, that later became the *Adriatic plate*.

The opening of the Penninic Ocean caused space problems in central Europe. The Meliata-Hallstatt Ocean in the southeast stopped extending and began to be shortened again. Unfortunately, there are practically no relics of this ocean found in the Alps and therefore we know little about its fate. Most of our evidence for this ocean comes from the Balkan region and the Carpathians.

On the Adriatic plate itself, the opening of the Penninic Ocean triggered a complicated tectonic evolution which is best known from the structural geology of the Calcareous Alps: along the northern margin of the Adriatic plate a lot of lateral displacement occurred and substantial topographic relief developed. Rocks were stacked and Jurassic coral reefs grew on these high points. At the same time local deep ocean basins developed into which huge debris fans (olistolites) and deep sea sediments were deposited.

In the Austrian Salzkammergut region, deep sea sediments and shallow water sediments still lie next to each other. North of the Penninic Ocean – on the stable southern margin of the European continent quiet sedimentation continued, similar in style to the sedimentation that occurred throughout the

Above: The Bergünerstöcke (from left to right: Piz Mitgel, Tinzenhorn, Piz Ela) in Swiss canton Grisons are made up of Hauptdolomit of the Ela Nappe (Lower Austro-Alpine). This nappe is pushed over the top of the Jurassic and Cretaceous layers of the Err-Carungas Nappe on the level of the scree fans (with Raibl Formation below). The Err Nappe forms most of the rocks where the green meadows are and also the mountain between Tinzenhorn and Ela: Pizza Grossa. Further to the right there is the crystalline basement of the Err Nappe forming Piz d'Err (the black summit with the snow field) and Piz Calderas (the pointy summit to its right).

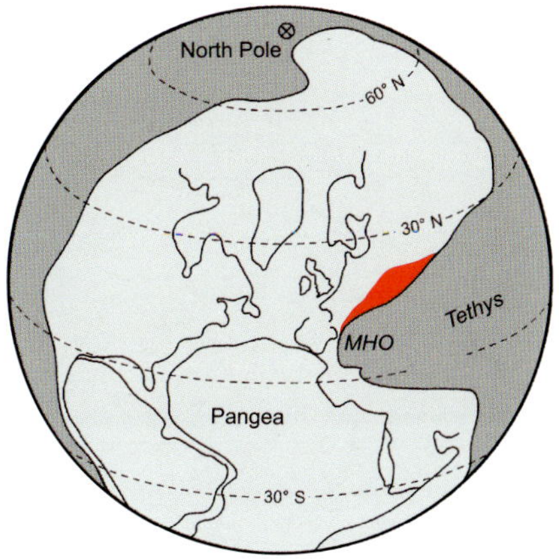

Late Permian (ca. 250 Ma)

Triassic on the Adriatic plate. The quiet sedimentation of calcareous shallow water sediments continued throughout the Jurassic and Cretaceous until about 60 million years ago. Later these sediments were folded and stacked into the Helvetic nappes of Switzerland.

At the turn of the Jurassic towards the Cretaceous, the Adriatic plate finally separated from the African plate and now formed a microplate of its own (often called the *Apulian plate*). At the same time Africa started drifting northwards and the convergence between African and European plates began that is still ongoing today. This convergence eventually led to the formation of the Alps as a mountain range. The oldest record of this *Alpine mountain building cycle* are thrust zones within the Austro-Alpine that formed around 135 Ma. These zones provide evidence that the northern part of the Adriatic plate was thrust underneath the southern part. The limestones that were earlier deposited in the Triassic and Jurassic were now scraped off successively from their basement (the northern part of the Adriatic plate) and stacked. The lower part of the crust submerged towards the south and suffered extremely high pressures and temperatures. Metamorphism led to the growth of new minerals (mica, garnet, staurolite, kyanite and many others) that form a large part of the crystalline basement of the Eastern Alps. 92 million years ago parts of these metamorphically changed rocks reached burial depths up to 80 kilometres. This is a depth normally only reached by rocks from the Earth's mantle. The old Permian gabbros were transformed to rocks with high density: *eclogites*. Not much later these eclogites were squeezed back up along the subduction channel. Enormous shear zones like the Stainzer Plattengneis in Styria (Austria) are the zones of motion along which this burial and exhumation occurred in the Cretaceous. Today these highest grade metamorphic rocks are found along a west-east oriented zone that leads from the southern Ötztal in Tyrol to the Schober Group, Kreuzeck Group, then to the Kor- and Saualpe and all the way to the *Sieggrabener Berg* along the easternmost part of the Alps. This line is possibly the trace of the suture zone along which the northern part of the Adriatic plate was thrust underneath the southern part. However the tectonics is not confined to this suture zone and most other rocks of the Eastern Alps were deformed during this time, for example the Wölzer- and Schladminger Tauern, Ötztal Alps and Silvretta.

From about 90 Ma onwards, the north-south directed nappe stacking was associated with extensional motion in east-west direction. A series of small marine basins formed on the surfaces that are called "Gosau basins", after the name of a location in the Salzkammergut region of Austria. The remnants of these Gosau basins may be found today on the Muttekopf near Landeck in Tyrol, in Gosau north of the Dachstein Massif and near Kainach west of Graz. The continuing convergence between Africa and Europe caused thrusting of the entire nappe stack on the Adriatic plate further and further northwards over the top of the rocks of the Penninic ocean floor.

The submergence of the Penninic Ocean sea floor underneath the thickened continental crust of the Adriatic plate formed a deep sea trench near the coast, not unlike the one located south of the Alaskan coast or west of South America today. This led to a change in the sedimentation pattern in the Penninic Ocean: deposition changed from the quiet, slow and passive sedimentation of deep sea sediments (*Bündnerschiefer*) to active deposition of sedimentary fans into the trench (*Flysch*). In the Eastern Alps, these sediments are called *Rhenodanubian Flysch* and form a long zone of rocks between the Northern Calcareous Alps and the Molasse Ba-

Below and right: Earth in the Jurassic and the Cretaceous. The red ring shows the approximate position of the Alps (after: McCann 2008).
P = Penninic Ocean
"A" = Adriatic appendage
A = Adriatic plate

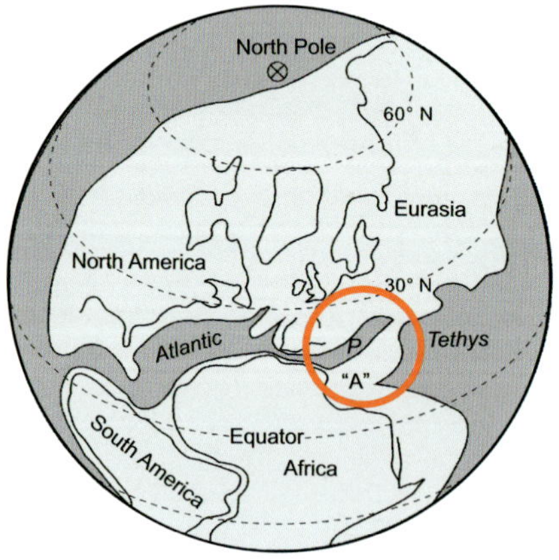

Late Jurassic (ca. 140 Ma)

sin. Similar rocks are also found in the Penninic schist hulls of the Tauern Window. In the Central and Western Alps flysch sediments are found in large parts of the Valais Alps and the Helvetic nappes.

During further convergence between the two plates, the crust of the Penninic ocean floor was pushed further and further underneath the Adriatic plate until the entire ocean disappeared, about 40 – 60 million years ago. The rocks that were dragged to depth experienced high pressure metamorphism. Blueschists and eclogites were formed along the southern margin of the Tauern Window and inside the Penninic nappes of the Central Alps. Once the entire Penninic ocean was gone, the continental European plate started to collide head on with the Adriatic plate. At the start, the southern rim of the European plate was dragged underneath the Adriatic plate. The calcareous sediments from the southern rim of the European plate were sheared off their base and started forming the first Helvetic nappes.

During the collision of the two continents the rate of deformation in the Alps began to decrease. The two plates began to interleave about 50 million years ago. A period of dramatic folding and nappe formation started – mainly in the Central and Western Alps. In the Central Alps the uppermost parts of the Adriatic plate were domed up and started eroding. As a consequence, there are barely any rocks from this upper plate left in the Central Alps. Only the rocks of the lower levels of the plate tectonic sandwich remain there: the rocks of the Penninic ocean floor and the fragments that lie as microcontinents within it (*Briançonnais-* and *Sesia* fragment). Even rocks from the deepest level of the sandwich (from the European plate) may be seen in places where erosion has reached deep enough: in the Internal Massifs, the External Massifs and the Zentralgneise of the Tauern Window.

The Internal Massifs (Sub-Penninic units) are fragments of European crust that were detached from their own base during the collision of the two continents and now lie like huge bubbles amidst the Penninic rocks: the Monte Rosa Massif, Gran Paradiso Massif and Dora-Maira Massif belong to these Internal Massifs.

At the same time another huge dome shaped structure of metamorphic rocks developed in the Central Alps: the *Lepontine Dome*. This dome was rapidly exhumed towards the surface along the Simplon Fault. All around this dome there is a zone of Penninic nappes and parts of the former European plate.

Somewhat further north, there are more dome-shaped structures: the External Massifs. These External Massifs are made of gneisses and granites of the Mesozoic European crust and include the Aare, Mt. Blanc, Pelvoux, Argentera and Belledonne Massifs. Around these External Massifs (and west and north of them) are the limestones and other sediments of the old European continental margin. They were partly sheared off their base, but partly remained intact. These are the Helvetic nappes of Switzerland that form some of the most spectacular fold patterns in the entire Alps.

Amid this dramatic evolution a new important line formed around 35 – 30 million years ago. In the middle of the Adriatic plate a deep seated fault zone cut across the entire plate: the *Periadriatic Line*. The suture zone starts in the Pohorje mountains of Slovenia, goes along the Gail Valley, Puster Valley and Valtellina to the west. At depth this structure now formed the separation line between the European and Adriatic plates. South of this line, the Adriatic plate was now thrust to the south instead of to the north. The southern Alps started developing as a tectonically separate part of the Alps. Soon after this line was formed, granites intruded the fault zone at

Above: The mountains of Valais and their most famous peak the Matterhorn. Geologically, the region is famous for its juxtaposition of rocks derived from the Mesozoic Penninic Ocean and rocks derived from continental fragments of the Adriatic plate and the Briançonnais Fragment. The nappe stack is also overlain by a klippe of Austro-Alpine rocks: the Dent Blanche Nappe.

Late Cretaceous (ca. 80 Ma)

around 33 Ma: the Bergell, Adamello, Rieserferner and Rensen granite. Only the easternmost of these intrusions, the Pohorje pluton, intruded a lot later and may have a separate story (around 19 Ma). The cause of formation of these granites is not very clear. However, most likely the melting process occurred in connection with the evolution of the subducted plate of the Penninic Ocean deep underneath the Alps (p. 196). This plate now hung underneath the Alps deep in the Earth's mantle. 35 – 30 million years ago this plate tore off – a process simply called by geologists: *slab break off*. The gap in the tearing plate suddenly provided both: pathways for melts to intrude the crust and the buoyancy for the Alps to rise. The Alps finally began to pop up like a cork from water. For the first time an Alpine mountain range started forming.

Central Europe can only be recognised as such from about 30 Ma onwards, as the Alps finally received a shape that resembles today's geometry. Nevertheless, the time of topographic development – the age of mountain building in the strict sense – is still debated. A long way east of the Alps, the geological development of the *Pannonian Basin* began around 25 million years ago: The giant, almost round, region between

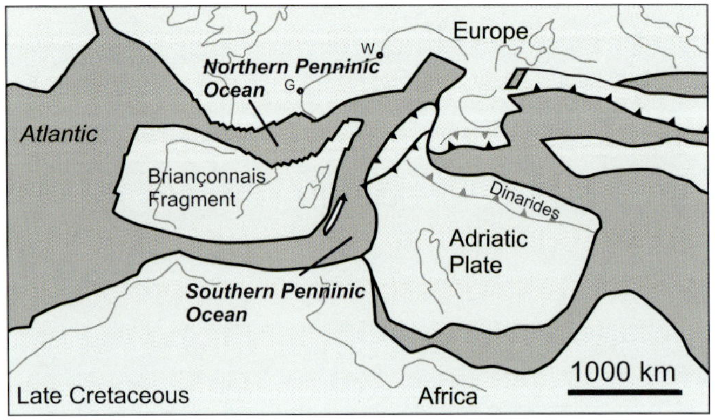

Above: The Brenta Group of the Southern Alps on the west side of Lago di Garda is part of a giant carbonate platform that formed in the Triassic and Jurassic on the continental shelf of the Adriatic plate.

Right and p. 91 bottom: The Earth in the Cretaceous and the Paleogene. The rectangular map is a cutout from the globe on p. 89 (after: Schmid et al. 2004).

the Carpathian arc and the Alps became a region where the lithosphere began to thin out and started to sink. Later the crust became so thin that volcanoes penetrated it. At the same time, the continuing collision between the Adriatic plate and the European plate squeezed the Alps eastwards into the region of thinned crust. In the Eastern Alps, this period between about 25 and 10 million years ago is called the: *"period of lateral extrusion"*. Just like toothpaste, the Alps were squeezed in a north-south direction and extruded eastwards towards the Pannonian Basin. Regions like the Wienerwald, the Styrian ranges Gleinalpe, Koralpe and Fischbach Alps, and the Pohorje Massif of Slovenia are like fingers that were then pushed towards the basin. The extrusion occurred along huge strike slip faults: the fault zones along the Salzach, the Enns, the Inn, the Gail, Möll, Mur, Lavant and Drava are the lines of motion for this extrusion tectonics. Simultaneously, north-south oriented detachments formed that opened the large tectonic windows. The upper nappes of the former Adriatic plate were pulled to the east and west and rocks similar to those of the Central and Western Alps were exposed: the rocks of the Penninic ocean floor were brought to the surface from underneath the continental rocks of the former Adriatic plate. We know of three tectonic windows where these Penninic rocks are seen through frames of nappes of the upper plate: the Engadine Window, the Tauern Window and the Rechnitz Window in the very east. The largest of these three windows, the Tauern Window, was pulled open along two spectacular detachment surfaces: the Brenner Fault in the west and the Katschberg Fault in the east. In all other parts of the Eastern Alps the higher-lying parts of the Alpine nappe pile (the Mesozoic Adriatic plate) are still intact because the east-west motions were compensated predominantly by strike slip motion.

In the Central and Western Alps, there was not so much room

to accommodate the north-south shortening by east-west extension. As a consequence, dramatic folding continued there throughout the Miocene, as it has kept going since about 40 – 50 Ma. In the region of the highest mountains of the Central Alps, the External Massifs continued to be domed up. On the northern slopes of these massifs, the Helvetic nappes suffered more and more deformation. The topography of the Alps itself finally began to lift up and their weight bent the European plate downwards. The Molasse basins formed north and south of the Alps as a consequence. These basins were rapidly filled with kilometre-thick sequences of sediments. All the way into the latest Miocene, these so called foreland basins still record marine-derived sediments. Simultaneously with this evolution, the region of the most intense shortening between the European and the Adriatic plates moved successively outwards from the central parts of the range to the northern and southern Alpine forelands. Finally, the Alps began to be thrust over their own foreland basins towards the north and south. At some stage, the areas of shortening jumped even further north, so that the Swiss Jura began to fold 20 million years ago.

About 7 – 10 million years ago the entire Alps, plus the surrounding basins and forelands, began a slow static uplift without any obvious signs of tectonics. For the Eastern Alps, geologists associate this uplift with the disappearance of an ocean in the region of the Carpathians. However, the uplift is so wide-spread and so lacking in any accompanying tectonic motions, that geologists can merely guess at its cause. The ocean that filled the Pannonian Basin became brackish, until the remaining freshwater lake finally dried up about four million years ago. The Molasse basins north and south of the Alps were also uplifted and partially eroded, giving the region south of Munich or north of Milan its hilly appearance. Even the Alps themselves were uplifted a bit more during this recent part of the evolution. Gravel deposits on the summit plateaux of the Calcareous Alps ("Augensteine") give evidence that the easternmost part of the Alps was uplifted by almost 1000 m within the last five million years. At the same time the sea level in the Mediterranean Sea sank intermittently by more than one kilometre: 5.3 million years ago the straits of Gibraltar closed tectonically for about 500000 years and the Mediterranean Sea dried up. The large Italian lakes – Garda, Como, Lago Maggiore – were carved by rivers that then dropped into the almost dried up Adriatic sea basin. Even today there are salt deposits many hundreds of metres thick on the floor of large parts of the Mediterranean Sea, which formed by evaporation of the sea at that time.

A massive change in climate two million years ago triggered the glaciation periods and literally gave the Alps their final polish. Most of the glaciation periods of Europe occurred within the last 600000 years of Earth's history. The large U-shaped valleys of the Rhône and Rhine were carved out and many of the smaller rivers were re-directed into new valleys. Today's drainage patterns in the Alps and the position of the drainage divides is therefore only a matter of the last few million years. Only the region east of Völkermarkt in the Austrian province Carinthia, east of Judenburg in Styria and the Napf region of Switzerland remained without glaciers throughout this time. These parts are today the only regions where geologists can study how the Alps may have looked geomorphologically before the glaciation periods.

Today, GPS measurements show that the Adriatic plate turns counterclockwise around a rotation pole near Turin in Italy and barely drifts northwards. Therefore, the Alps are tectonically quiet to the north of Turin. In the Tagliamento region north of Udine, the convergence between Italy and Europe is at its maximum: several millimetres per year. This region is

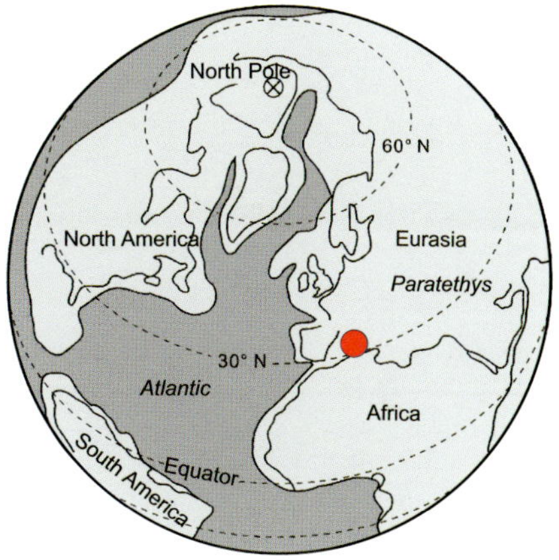

Paleogene (ca. 25 Ma)

Block diagram showing the evolution of the Alps from the Upper Triassic (about 215 Ma) until the Oligocene (30 Ma) (after: Mandl & Brüggemann-Leodolter © GBA). In the Triassic (215 Ma), huge carbonate platforms formed on the southern margin of the Adriatic continental plate (labelled here as "Alpine Shelf") at the northern rim of the Tethys Ocean (Meliata-Hallstatt Ocean). Some time between the Upper Jurassic and the Lower Cretaceous (140 Ma), the Tethys Ocean was displaced eastwards by gigantic strike slip zones. A new ocean had opened up further north: the Penninic Ocean. During the upper Cretaceous and the Eocene (60 Ma) the south-directed subduction of the Penninic Ocean was in full swing. In the Oligocene this ocean was gone and the head-on collision between the Adriatic plate and Europe dominated Alpine tectonics.

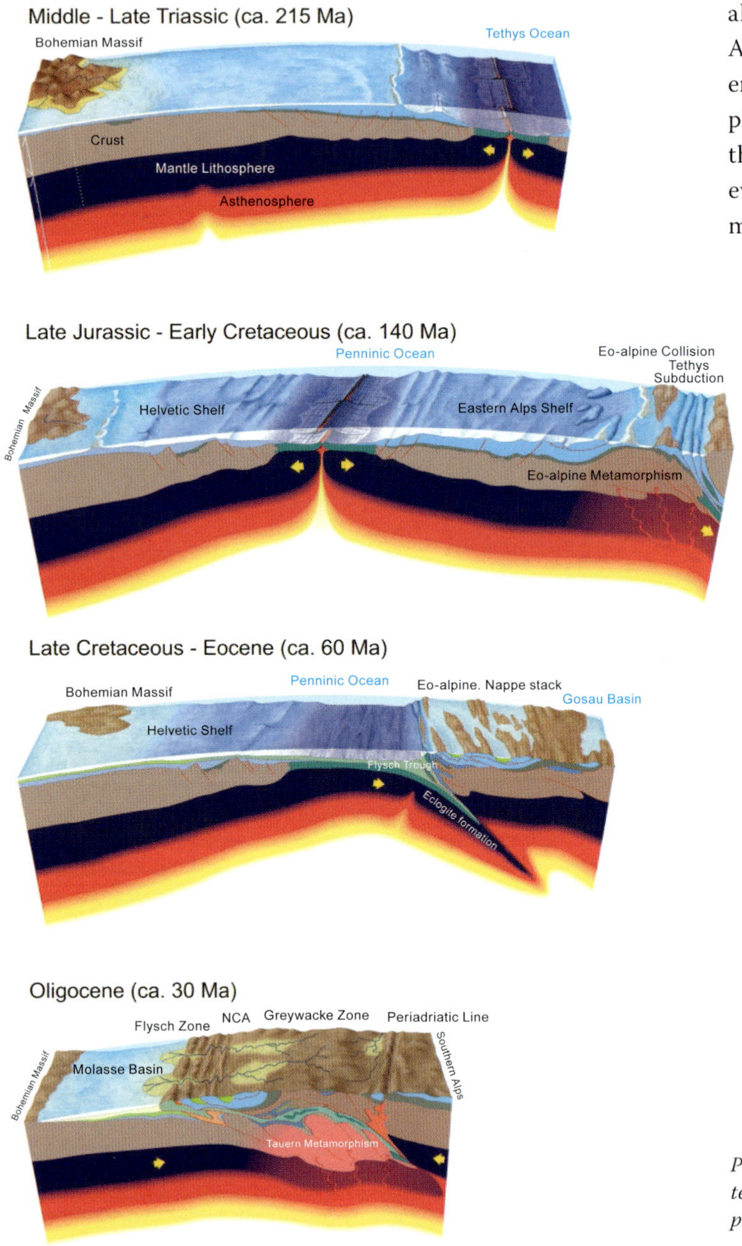

also the most seismically active part of the Alps. Uplift of the Alps continues at several tenths of millimetres per year, but erosion by rivers holds the pace. The rotation of the Adriatic plate turns all of Italy and drags Sardinia and Corsica out of the southern rim of France with astonishing speed. This, however, is a story that goes far beyond the geology of the Alps and must be told some other time …

Page 93: Autumn in the lower Adige Valley. The mountain with the telecommunication towers is Mt. Paganella (2 125 m), a well known view point and ski resort with a view of Lago di Garda.

3.1. “Stone Age” of the Alps

Page 95: Cellon (2241 m) on the west side of the Plöcken Pass in the Carnic Alps is famous for its Paleozoic stratigraphy and for the oldest fossils found in the Alps. The famous Wolayer Lake lies directly behind the summit of Cellon. The summit in the clouds is called Hohe Warte (2780 m).

Below: Stratigraphic profile illustrating part of the sequence seen at Cellon. The photo to the right of the profile shows the Carnic Alps with the Hohe Warte seen from the west.

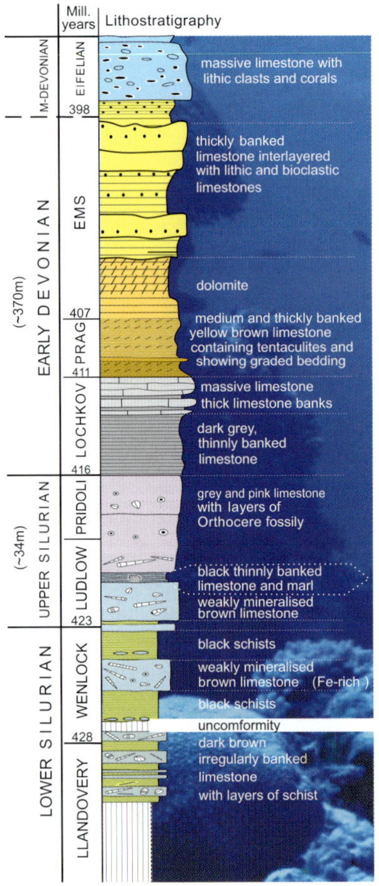

A. Remnants of Gondwana

Remnants of Paleozoic Gondwana are only known in the Eastern Alps: Ordovician and Silurian sediments (490–416 Ma) that were once laid down into an ocean along the northern margin of Gondwana. Rocks of such Paleozoic age are known from five locations in the Eastern Alps:

1. in the Paleozoic of Graz in the Austrian province Styria
2. in the Gurktal Nappe between the rivers Gurk and Mur
3. in the Greywacke Zone
4. much of the Carnic Alps
5. in a small part of the Karavanke mountains.

In these five regions ancient limestones and schists occur that are partially fossil bearing and have not been overprinted metamorphically ever since. These rocks can therefore be used to reconstruct the evolution of life and climate during this early age of Earth. The Cellonrinne at the base of the 2241 m high Cellon (Frischenkofel) west of the Plöcken Pass in the Carnic Alps is considered as a unique standard for the sedimentary evolution of Europe in the Silurian. Between 450 and 400 million years ago differently coloured limestones and schists were deposited over a period of almost 50 million years. Trilobite fossils in the 60 m thick sequence show that the rocks were deposited into a shallow sea. The section in the Cellonrinne is world famous for its undisturbed layering and fossil wealth. It was discovered at the end of the 19th century and has been studied ever since. Above the famous Silurian sequence there are other limestones of Devonian age that make up the upper parts of the Cellon mountain.

Hohe Warte (2780 m)
Wolayer Lake
Mooskofel (2506 m)
Cellon
(2241 m)
Cellon limestone
(380-390 Ma)
Vinz limestone (405 Ma)
Kellerwand limestone
(410 Ma)
Boden l. (415 Ma)
Rauchkofel l.
(>420 Ma)

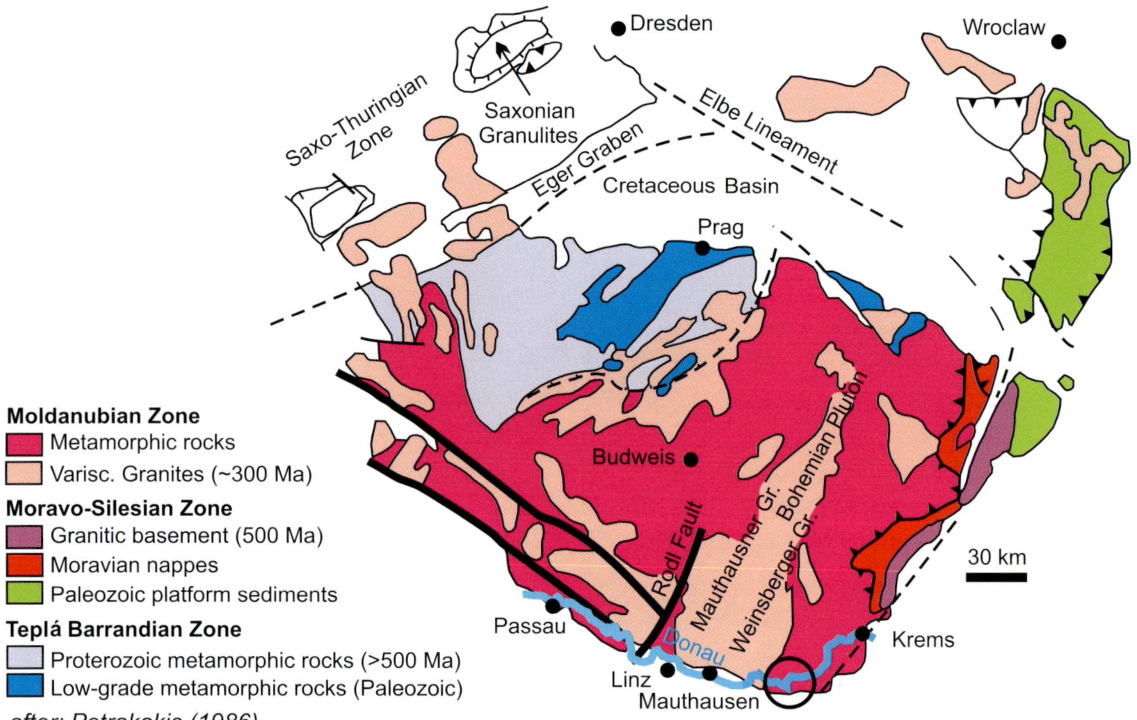

after: Petrakakis (1986)

Above: Geological map of the Bohemian Massif (ring = photo, p. 97).

Right page: The Benedictine monastery of Melk in Lower Austria was founded by the Babenberg dynasty in the 11th century. Ever since it has guarded the entrance to the Wachau region that, in this photograph, offers an impressive example for recent tectonic activity in some of the oldest rocks of the Alps: here, the Danube leaves the flat plains where it flows through the soft Molasse Zone sediments. Instead it digs into the hilly landscape of the hard Variscan gneisses of the Bohemian Massif. The Schönbühel monastery in the background stands on the contact between the Molasse Zone and the Bohemian Massif.

B. Variscan Rocks

The Variscan orogeny (380 – 310 Ma) in the Alps is best explained using a region that actually lies outside the Alps: the Bohemian Massif. Geographically, the Bohemian Massif includes the Lower and Upper Austrian provinces north of the Danube and reaches all the way to the Saxonian Erzgebirge. All this is part of the Variscan Europe that builds up large parts of Europe north of the Alps. The rocks of the Bohemian Massif are Paleozoic and even Precambrian granites, gneisses and sediments that were heated up around 320 – 350 Ma in the lower crust and thus transformed to high grade metamorphic rocks.

Geologically, the Bohemian Massif may be subdivided into three major parts: the *Tepla-Barrandian Zone* to the north consists of Precambrian rocks of relatively low metamorphic grade.

The *Moldanubian Zone* builds up most of the massif and is a typical representative of the Variscan rocks. It is composed of migmatitic gneisses that can be found along the valley of the Danube. Finally, in the very east of the Bohemian Massif, there is the *Moravic Zone*. This zone was overthrust in the high phase of the Variscan orogeny (in the Carboniferous) by the Moldanubian Zone. At the end of the Variscan orogeny about 300 Ma large volumes of granite (e.g. Mauthausener granite, Weinsberger granite) intruded the metamorphic basement. Afterwards, the Variscan mountains eroded, the Bohemian Massif was levelled out and covered by terrestrial sediments.

Many of the Variscan gneisses of the External Massifs and the Central gneisses of the Tauern Window experienced a similar history, which cannot be observed anymore, because they were later incorporated into the Alpine tectonic processes. In contrast, the Bohemian Massif was left completely untouched by tectonic processes since about 290 Ma, and was finally slowly uplifted to its present elevation of about 500 – 700 m in the very last few Ma. Plöckenstein near the Czech – Austrian border is even 1 378 m high. The age of this recent uplift is unknown, but the course of the Danube gives an indication: the river changes its course in two locations from the flat landscape of the soft Molasse Zone rocks into the much harder rocks of the mountainous landscape of the Bohemian Massif: between Passau and Linz and again between Melk and Krems (Wachau region). On both occasions, the river does not follow the wide plains of the Lower Austrian Molasse Basin, but flows towards the mountains and into a steep sided valley inside the Bohemian Massif. From this, geologists interpret that the Danube is much older than the uplift of the massif. Therefore, the uplift cannot be older than 10 – 20 Ma at most.

3.2. The Permian: Some Like It Hot

Page 99: The valley of the river Toce cuts across the rocks of the Ivrea Zone and connects the north Italian city of Domodossola and the Lago Maggiore. The profile through the Permian crust along its sides is the most famous of its kind in the entire Alps.

Right: The bluffs on the broad ridges of the Saualpe crest are made up of eclogites that formed in the Cretaceous from Permian gabbros. Their high resistance to weathering makes them stand out of the landscape.

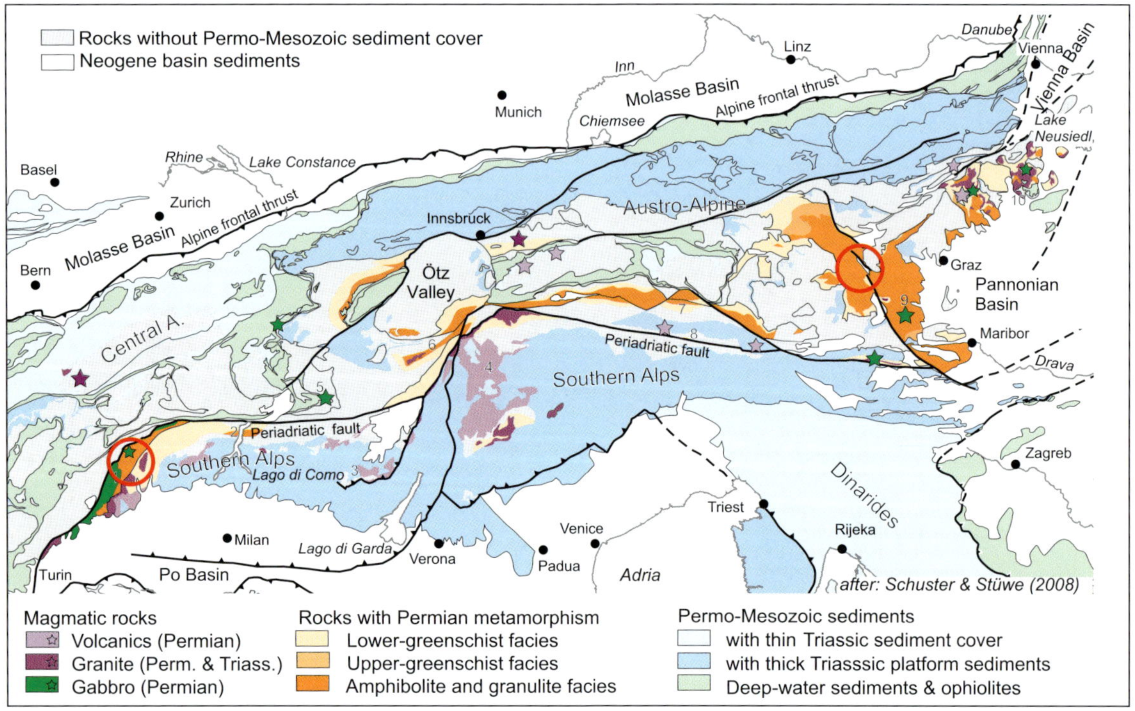

Below: Map showing the distribution of Permian rocks and metamorphism in the Alps. West of the shown part of the Alps there is no Permian influence (ring at left = photo, p. 99; ring at right = photo, p. 98).

During the Permian (251 – 299 Ma) volcanoes erupted, granites intruded the crust and rocks were metamorphosed at high temperatures. In short, it was a hot time. Interestingly, there is no widespread deformation event that accompanies this thermal event. The Ivrea Zone northwest of Lago Maggiore is chosen here as an example for the evolution of the Alps at this time, because it records a unique cross section through the crust as it formed in the Permian. The region was tilted on its side after the Permian events and the river Toce carved its valley through the region in the most recent part of its geological history. Today the steeply dipping rocks along the Toce Valley allow a journey across the entire Earth's crust. Near Vogona, rocks from the Earth's mantle are exposed. The further one drives from these towards Lago Maggiore, the higher the exposed crustal level: initially, the rocks are paragneisses and gabbros that were metamorphosed in the Permian to extremely high temperatures. Then the grade of metamorphism successively decreases from granulite to amphibolite facies (p. 84), until the contact with the Permian upper crust is reached near Montorfarno. Southeast of there the Strona-Ceneri Zone follows, which represents the Permian middle crust. The Ivrea Zone itself can be followed geophysically underneath the Po plains.

East of the Ivrea Zone, Permian rocks also occur in many other parts of the Alps. The *Bozen quartz porphyry* is one of the best known examples for Permian volcanics. It erupted from a huge caldera (several hundreds of square kilometres in size) that lies underneath the Dolomites. Permian metamorphics occur in the Eastern Alps along a narrow band south of the Tauern Window from the Schneebergerzug in the Ötztal Alps to the Saualpe (2 044 m) and Koralpe (2 140 m) in Styria. Gabbros that intruded the crust in the Permian were later metamorphosed to eclogites (p. 104). Tourmaline, spodumen and beryl bearing pegmatites in the region are also of Permian origin.

Strona-
Ceneri Z.
Lago Maggiore
Baveno
Granite
Montorfano
Granite
Ivrea Z.
Ivrea Z.
Sesia Zone

3.3. Triassic: The Time of the Coral Reefs

Right and below: The Kuhtrittmuschel from the family of the megalodontes is a typical fossil of the Triassic. It is common in the Dachstein limestone (photo: J. Gudelius). The geological cross section through the Calcareous Alps on the level of Dachstein shows the thickness of the Triassic carbonate sequences.

Page 101: The Dachstein south face in the Austrian province of Styria is a classic example of the carbonate development in the Triassic. It is also the type locality for the Dachstein limestone that occurs in many parts of the Alps. On the level of the Ramsau village, the carbonates lie on the schists of the Werfen Formation (250 Ma). The peak at right with the small cloud over its summit is the Kammspitze (2139 m). To the right of it is the Stoderzinken (2048 m), and to the left the impressive massif of the Grimming (2351 m). Behind Grimming the mountains of the Gesäuse may be seen. The Koppenkarstein (2865 m) lies in front. Left of it, the top end of the cable car onto the Dachstein. The three main summits towering above the Dachstein south face are (from right to left): the Hohe Dachstein (2995 m), Mitterspitz (2925 m) and Torstein (2948 m). The mountains of the Tote Gebirge lie on the sky line.

In the Triassic (199 – 251 Ma) more rocks were formed in the Alps than in any other period of Earth's history. Most of these newly created rocks are carbonates that formed as giant platforms along the margins of the Tethys Ocean (*Meliata-Hallstatt Ocean*). What was to become the *Northern Calcareous Alps* and the *Dolomites* grew in the Triassic along the southern margin of this ocean. The shallow water carbonates that formed along the northern margin of the same ocean later became the *Helvetic nappes of Switzerland* and the limestones of southern France. Overall, the Tethys region at this time may have looked a bit like the Carribbean today. Carbonate platforms like the Bahamas filled a shallow tropical ocean that was underlain by the southern rim of the European continental plate. During all this carbonate formation, the tectonic extension that once affected the surrounding continents continued way out in the middle of this ocean.

The marginal facies of some of these coral reefs are still preserved today: on the *Steinplatte* (p. 170), in the *Hochschwab region* (p. 142) and at several places in the Dolomites. The fossil content of these Triassic carbonates shows the geologist that many of these rocks were deposited in light-penetrated waters. As these carbonates are often several kilometres thick, this observation implies that the con-

tinental margins at the northern and southern end of the Tethys Ocean must have sunk successively deeper down. Geologists call such a tectonic scenario "sag phase". Sag phases are characteristic for a time of tectonic quiescence at the end of extension cycles. They are caused by slow cooling of the lithosphere. The thick limestone sequences of the Dachstein south face in the Austrian province of Styria are a spectacular document of this evolution.

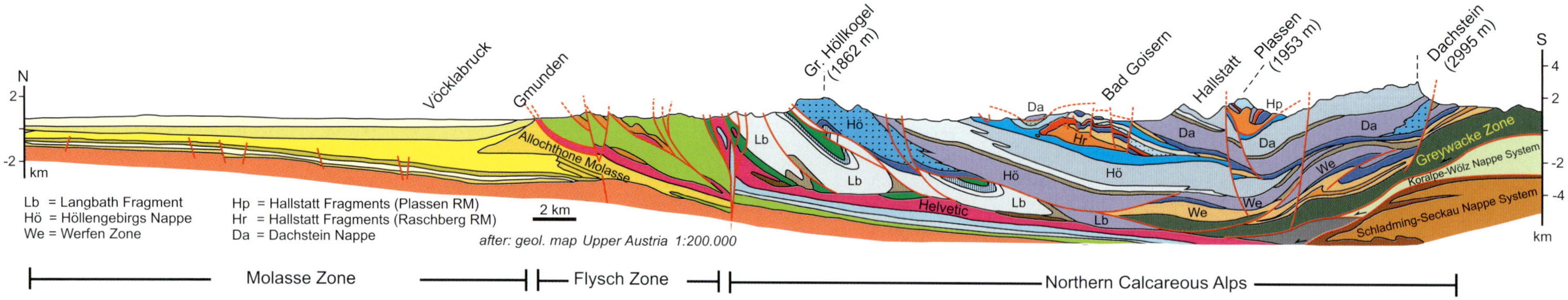

3.4. Jurassic: An Ocean is Born

Right: The pillow-like structure of the volcanic rocks in this photo indicates that the lava once flowed under water (photo: G. Pestal). This example is from rocks that belonged to the Jurassic Penninic Ocean.

Below: Block diagram of the Central Alps showing the rocks of the Penninic Ocean floor in green, Briançonnais in orange, South Alpine and Austro-Alpine (AA) in red and parts of Europe in light brown. Sesia Zone and Dent Blanche Nappe (DB) in yellow (after: Steck, coloured by J. Pleuger), ring = photo, p. 103).

Page 103: Grand Combin high above Lac de Mauvosin lends its name to the geological Combin Zone, a narrow band of rocks on the south side of the Rhône Valley that is made up of rocks of the former ocean floor of the southern Penninic Ocean. The rocks include remnants of ocean derived lavas and deep sea sediments that were later transformed into calcareous mica schists.

The Jurassic (199 – 145 Ma) was characterised by the opening of the Penninic Oceans. These oceans formed roughly in the region that is occupied by the Alps today and began to separate the Adriatic from the European plate in the Jurassic. The north-south extension that opened this ocean also meant a re-start of the tectonic processes in the flanking continents. Initially, the Earth's mantle was pulled out from underneath the northern margin of the Adriatic plate. The Triassic reefs were pulled into pieces and small deep ocean basins lay next to shallow marine regions. We know this from the juxtaposition of deep sea sediments (radiolarites) and Jurassic shallow water carbonates, for example in the Salzkammergut region of Austria.

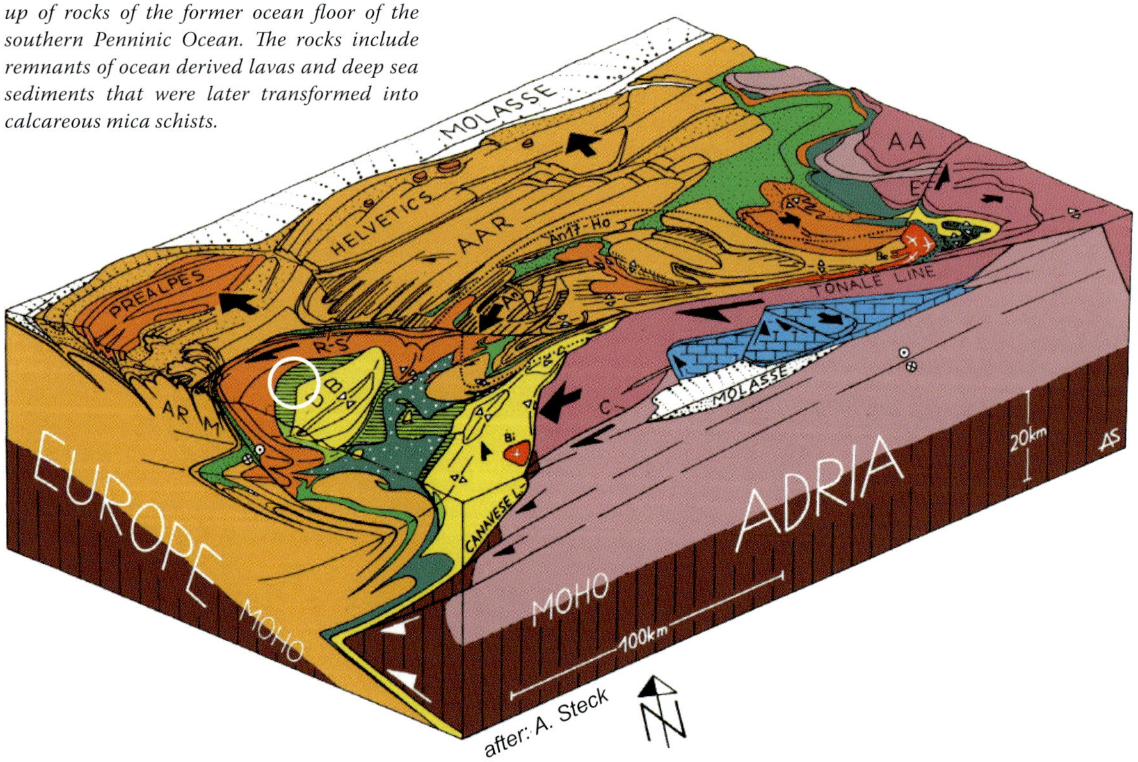

The record of this young Penninic Ocean is given by oceanic rocks of Jurassic age that are found across the Alps. In particular, these include:

1. *Sediments,* that were deposited in the middle of this young ocean (*Bündnerschiefer*).
2. *Basaltic rocks* of the ocean floor itself.
3. *Shelf sedimentation* on the continental margins (later Helvetic zones).

Despite later metamorphism, pillow lavas are still recognisable in rare locations in several places in the Alps. Such pillow-shaped structures form when lava extrudes under water and cools very rapidly on its surface.

The *Combin Zone* – named after the mountain Grand Combin (4 314 m) – is a typical representative of the Jurassic rocks of the Piemont-Ligurian Ocean. This ocean was the southernmost and first-formed branch of the system of Penninic oceans. The Combin Zone leads from the southern side of the Rhône Valley through Grand Combin and then as a narrow band of rocks all the way into the southern part of the French Alps.

3.5. Cretaceous: The Heyday of Early Alpine Tectonics

Page 105: The sleepy landscape of the Sau- and Koralpe in the Austrian provinces Styria and Carinthia makes it hard to believe that the region was the centre of tectonic activity in the Alps around 90 million years ago during the Eo-Alpine event. Just before this time the rocks were dragged down to a depth of about 80 kilometres and transformed to garnet-bearing gneisses, eclogites and marble at temperatures around 700°C.

Below and right: The crest of the Saualpe looking northeast towards Gleinalpe and Ameringkogel (2 187 m). In the right background the Mur Valley. The Koralpe range lies just out of the picture to the right. The rocky crags in the foreground are locally called "Felsöfen". They are made from the weather resistant eclogites. The spectacular hand specimen at right is made from kyanite (blue), garnet (red) and omphacite (green) and was collected in the Pohorje Massif just south of the Saualpe.

In the Cretaceous (145 – 65 Ma), the crust of the Penninic Ocean floor was successively subducted towards the south underneath the Adriatic plate. Some carbonates were still deposited along the northern shores of this ocean. These built up what are today the uppermost layers of the Helvetic nappes of Switzerland. However, the most important event of this time is the peak of the subduction activity and the associated tectonic and metamoprhic events: the *Eo-Alpine event*. Rocks were dragged down to extreme depths and transformed to garnet mica schists that constitute the bulk of the Austro-Alpine basement. The highest grade metamorphic rocks lie along a belt of rocks between *Schneebergerzug* in the west, *Kreuzeck Group* and *Koralpe* in the east. This line marks the trace of the subduction zone at this time. Today only few individual bands of motion from this subduction zone are visible. The shear zone of the Stainzer Plattengneis in Styria of

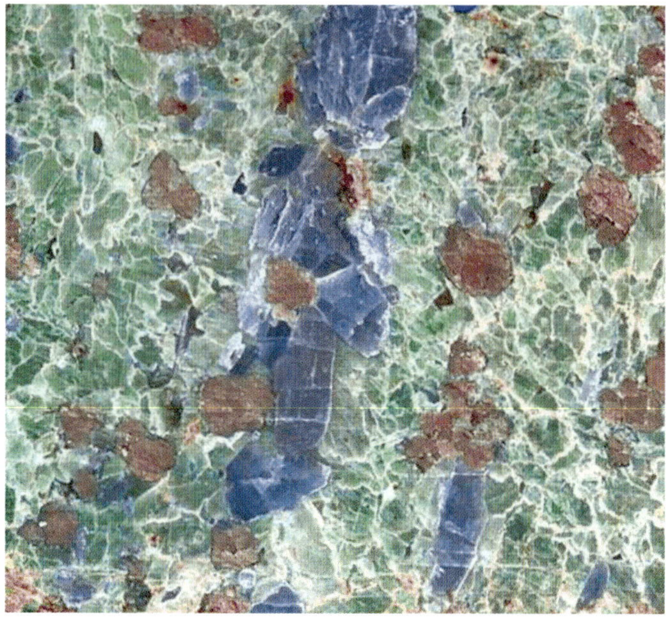

Austria is one of these. About 70 Ma ago these processes were over. Extension and erosion at the surface began to exhume the intensely deformed and metamorphosed rocks. In the Kor- and Saualpe region (Styria, Austria), which served here as the example for the Eo-Alpine metamorphic event, rocks reached the surface at around 40 million years ago. However, the surface at this time was flat and the Alps did not exist yet as a mountain range.

Today's topography of the Kor- and Saualpe is the consequence of a very recent uplift event: only about 5 Ma ago the surface began to rise again. The unusual wide rolling landforms high above 2 000 m sea level testify to this recent uplift event: rivers have not found time yet to dissect the plateaus. Moreover, the Kor- and Saualpe were never covered by ice during the glaciation periods, so that glacial cirques are absent.

3.6. Paleo-/Neogene: The Alps Begin to Shape Up

The Simplon detachment at the Simplon Pass between Switzerland and Italy is one of the large Neogene structures of the Alps. It is responsible for the exhumation of the Lepontine Dome (see also p. 250) The western slopes of Punta Giezzo (2658 m) and Pizzo Pioltone (2610 m) are the detachment surface itself (on the sketch in red). From there, the eastern side of the fault zone was pulled out many kilometres from underneath the western side. For more information on Simplon Fault see p. 250.

The Paleogene (65 – 23 Ma) and Neogene (23 – 0 Ma) include the time period since the mass extinction of many species including the dinosaurs at the end of the Cretaceous. The two periods together replace the term Tertiary in the modern division of the geological time table. During the early Paleogene the tectonic processes of the Cretaceous continued: the last bits of the Penninic Ocean continued to subduct in the west of the Alps. Around 20 – 30 Ma the Penninic Ocean was gone completely. The geographic region of the Alps slowly adopted its shape and geometry as we know it today. From then on it becomes easier to follow the geological evolution, because it is only then that mountains and structures formed in about the geographic position where they still are today. The mountain building in the narrow sense of surface uplift also only began at this time. The most important tectonic events of the last 30 Ma were:

A. *Large fault zones form.* This includes many of the shallow

nappe boundaries in the Central and Western Alps, but also the steep strike-slip zones of the Eastern Alps and the Periadriatic Line itself, the largest fault zone in the Alps.

B. *Granites intrude the crust.* The only granitic intrusions associated with the Alpine orogenic cycle intrude the Periadriatic lineament at around 30 Ma.

C. *Basement domes rise to the surface.* This concerns the tectonic windows of the Eastern Alps, but also the basement upwellings in the Central and Western Alps: the Internal and External Massifs and the Lepontine Dome. Some of these domes were exhumed along huge detachment zones, others remained at least in partial contact with their former cover rocks. The Rhône-Simplon Fault is the most famous of these detachment zones and was responsible for the exhumation of the Lepontine Dome. Parallel to the formation of these domes, large parts of the Central Alps were intensely folded.

Sketch above: The large strike slip fault zones at the eastern end of the Alps. Pull-apart basins (shown in yellow) lie along many of these faults. F = Fohnsdorf Basin, south of there the Obdach and Wolfsberg Basins. Between W = Vienna Basin and the Fohnsdorf Basin lie the Leoben-, Krieglach- and Parschlug Basins.

A. Large Fault Zones Develop

Among the large fault zones, shear zones and nappe boundaries that developed in the Alpine realm within the last 30 Ma, the steep strike slip zones of the Eastern Alps are the most impressive ones. Along them, the Alps were extruded eastwards, the block between the *Salzachtal-Ennstal-Mariazell-Puchberg-Fault* (SEMP) and the *Mur-Mürz-Fault* predominantly to the north and the *Periadriatic Line* and the *Lavanttal Fault* in the south. Several small pull-apart basins were ripped up and today bear kilometre-thick sequences of sediments. The photo shows the southern margin of the Wolfsberg Basin in the Lavant Valley as an example for the gigantic fault zones that were active at this time in many places across the Alps.

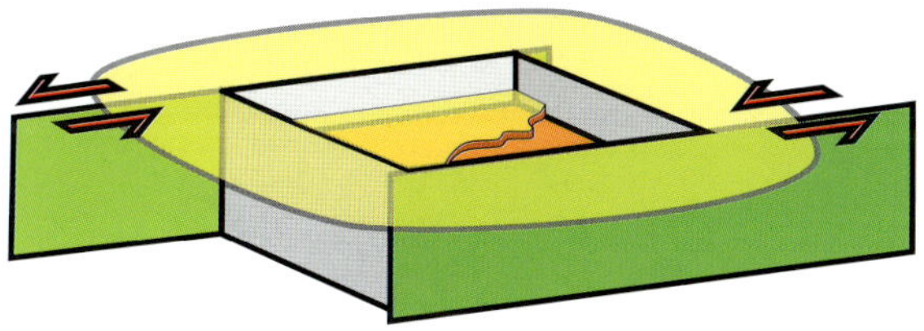

Above: The southern end of the Lavant Valley along which the Lavanttal Fault runs in the Austrian province of Carinthia. The section between Lavamünd and Dravograd (under the clouds in the middle ground) is shown. There the river Drava leaves the broad valley and makes a sharp kink towards the east into the narrow gorge of the Drava Valley within the Pohorje Massif. The mountains on the horizon are the Steiner Alps (Kamnisko Savinjske) at the eastern end of the Karavanke range. Their highest summit is the Grintavec (2558 m), seen at the right end of the range. The broad ridge in front of Grintavec is the Feistritzer Spitze (in German: Petzen, 2113 m).

Sketch at left: schematic sketch showing the formation of pull-apart basins. Left lateral slip along several parallel fault zones rips up a region in between, which then rapidly sinks and fills with sediments.

B. Granites Intrude the Crust

The only granites that are known to be associated with the Alpine orogenic cycle are those that formed along the Periadriatic Line. This structure formed a bit before 30 Ma amidst the Adriatic plate along an older scar zone. It is the longest fault zone of the Alps (see below) and has formed the boundary between the Adriatic and the European plates ever since. Since its formation, all rocks south of this line have been pushed southwards over the top of the even younger sediments of the southern Molasse Basin. When the structure formed at 30 – 34 Ma, granitoid melts from the lower crust intruded into it.

At least eight of these granitoids are known between Ivrea in the west and Maribor in Slovenia in the east. Only the easternmost of these granites, the Pohorje pluton, is significantly younger than all the other ones (19 Ma).

Because of the counterclockwise rotation of the Adriatic plate over the last 20 Ma, the northern side of the Periadriatic Line has been displaced many kilometres to the east relative to the southern side. A series of large valleys have incised along the shear zones that formed along this line. From east to west this line leads along the valleys of Drava, Gail and *Puster Valley*. The tectonic line then goes along the *Addatal (Veltlin)*, through the northern end of Lago di Como and Lago Maggiore and then towards the southwest all the way to Ivrea.

Right: Map of the Periadriatic Line and the granites that have intruded it. Aside from the Pohorje Pluton (19 Ma) in the east, all other granites are between 30 and 34 Ma old. A more detailed labelling of the various branches of the Periadriatic Line is found on p. 83.

Page 111: The Adamello Massif (foreground) in clouds. The massif is built up of the largest granite intruded into the Periadriatic Line (position of the photo is shown with the white ring on the map at right).

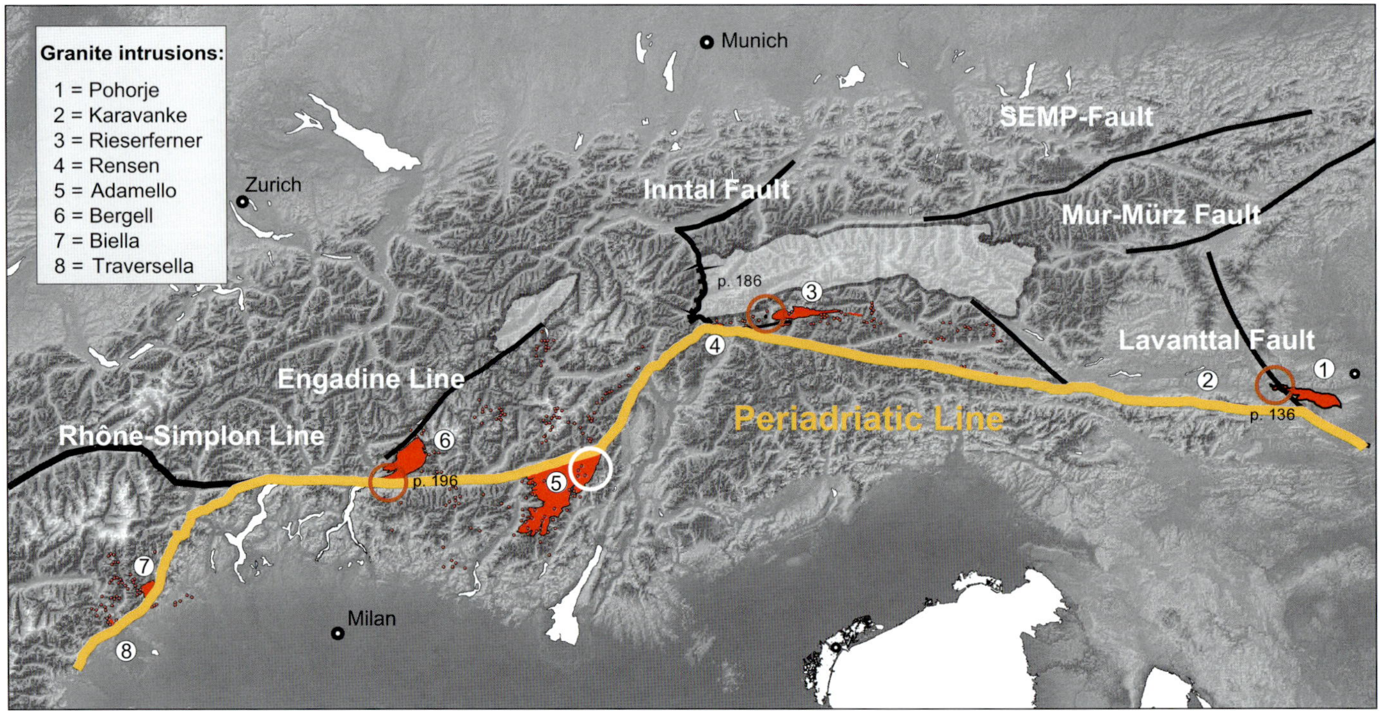

C. Basement Domes Rise

All across the Alps large domes of crystalline basement began to rise towards the surface at about 20 Ma. The External Massifs in the Central and Western Alps domed up but remained mostly in contact with their sedimentary cover in the process. In contrast, the tectonic windows of the Eastern Alps were literally torn open. The Brenner detachment along the western side of the Tauern Window is one of the most prominent of the large north-south oriented fault zones that tore open the window (p. 74, 158). The fault zone dips at about 20° to the west and mostly follows the Brenner Valley. Along this fault zone, the Tauern Window is continuously being pulled out towards the east from below the Austro-Alpine nappes. Earthquakes in the region show that the zone is still active.

Geologists know that the distance between the Hohe Tauern range and the Ötztal Alps has increased by more than 60 km during this process. Within the Tauern Window (east of the Brenner Pass), garnets and micas may be found that formed 20 Ma at more than 500°C, deep in the crust. West of the pass there are low grade metamorphic quartz phyllites of the Mesozoic Steinach Nappe and the dolomites of the *Tribulaun* (3097 m) and *Serles* (2718 m).

The Brenner detachment seen from the Brenner Pass looking north. The shear zone lies on the large wooded slopes that are inclined to the left. The Venner Valley (at right) leads into the inner parts of the Tauern Window. Not far from the Kraxentrager (2999 m) at the end of the valley the Central Gneisses start. At left Patscherkofel near Innsbruck may be seen in the background; behind it the Nordkette of Innsbruck. The block diagram illustrates the tectonic situation of the Brenner detachment (grey hatched region). The Tauern Window (green) was pulled out from there from underneath the Ötztal basement. The rectangle indicates the photo region.

3.7. Quaternary: The Final Polish

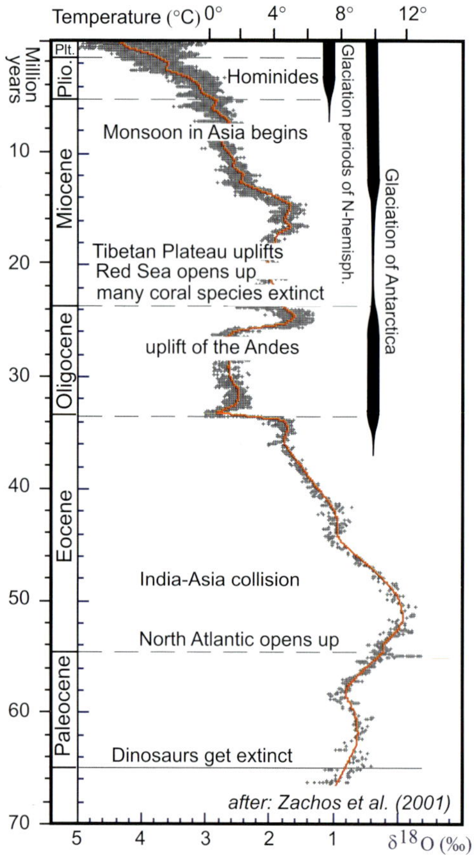

Above: Global climate curve for the last 70 Ma of Earth's history. The isotopic composition of oxygen ($\delta^{18}O$) in water depends on temperature. Therefore it is possible to use analyses of oxygen in fossil skeletons to infer climatically driven temperature changes.

Right: The Klafferkessel region in the Schladminger Tauern of Upper Styria in Austria. The glacial cirques and lakes are among the most famous of their kind in the Eastern Alps. Dachstein Massif in the background.

A. The Ice Ages

After a warm spell in the mid Miocene (at just less than 20 Ma), the global climate began to deteriorate. About 4 million years ago the temperature changes became more rapid and ice caps started forming in many places of the northern hemisphere. Large parts of North America were covered by the *Laurentide ice cap*. In Europe, ice fields covered all of Scandinavia and reached down to central Germany. The Alps had their own ice fields that remained unconnected to the northern European ice fields. However, this ice cap fluctuated substantially as it is now known that the last 650 000 years had a very complicated history with warm and cold periods. Nevertheless, many of the terminal moraines that we can see around the Alps today formed during the *Last Glacial Maximum* (LGM) around 19 000 – 25 000 years ago. 90% of the Alps were then glaciated and only the summits appeared as nunataks above the ice. Glaciers reached far into the Alpine forelands. About 20 000 years ago the Drava glacier still reached as far as Völkermarkt in Carinthia, where its terminal moraine may still be seen. The Mur glacier reached as far as Judenburg. The terminal moraine of the Tagliamento glacier is crossed by the freeway about 10 km north of Udine and the moraine of the Adige glacier about 10 km west of Verona. The Inn glacier covered all of Rosenheim, the Chiemsee and reached as far as Wasserburg on the Inn. The Rhine- and Rhône glaciers together covered much of the Swiss foreland and the region of todays cities of Geneva, Zurich and Bern. Many hundreds of metres thick layers of rock were scraped off the Alps in this time by the glaciers. After the end of the last ice age (about 11 700 years ago), a completely new landscape was born.

Page 115: Map of the Alps during the Last Glacial Maximum (after: J. Ehlers). The Napf region between the Rhône- and Rhine glaciers and the very east end of the Alps are the most important regions that remained ice free. Note that much of the northern Adriatic was dry at the time.

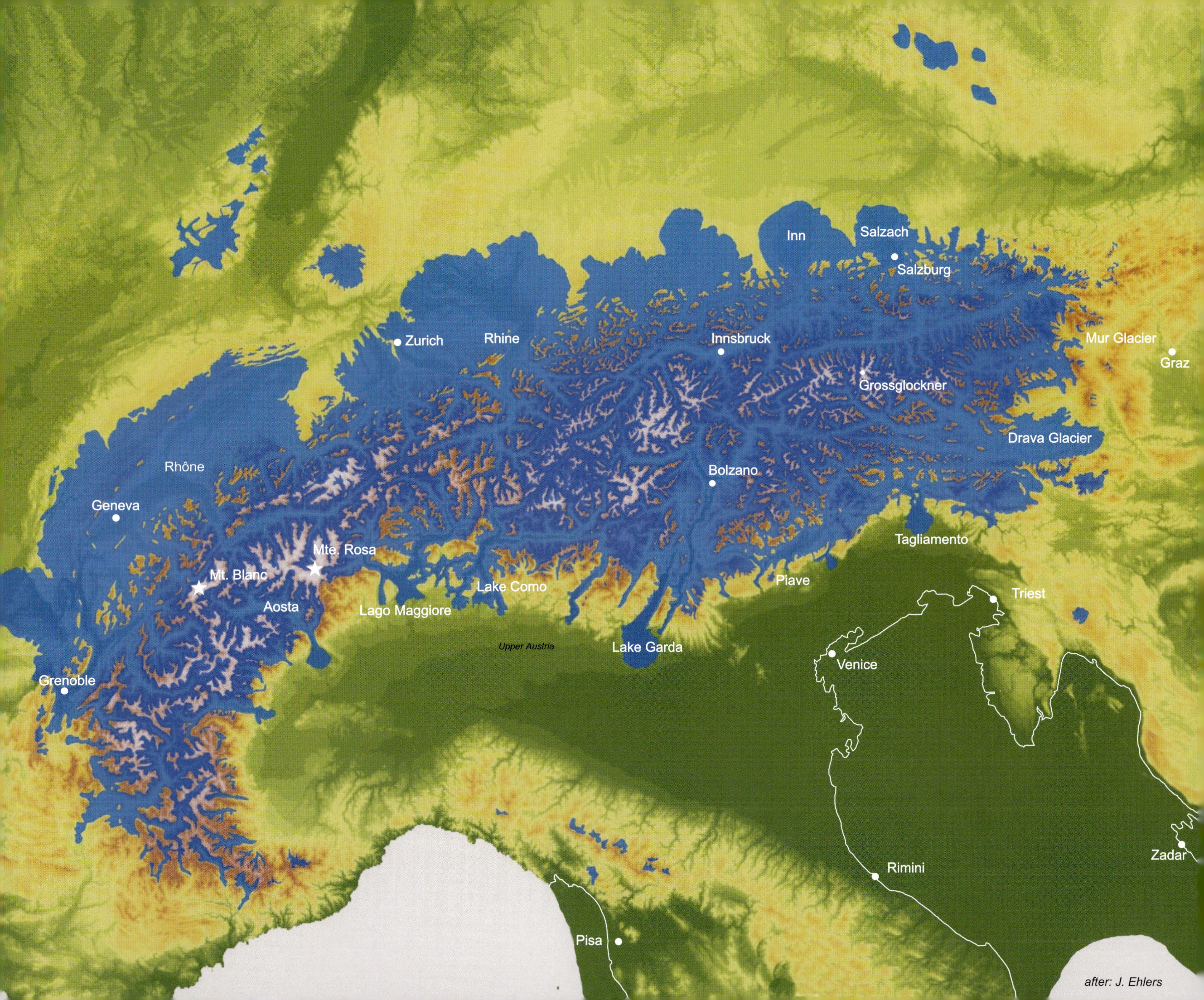
Inn
Salzach
Salzburg
Zurich
Rhine
Innsbruck
Mur Glacier
Graz
Grossglockner
Drava Glacier
Rhône
Bolzano
Geneva
Tagliamento
Mte. Rosa
Mt. Blanc
Lake Como
Piave
Triest
Aosta
Lago Maggiore
Upper Austria
Lake Garda
Venice
Grenoble
Rimini
Zadar
Pisa
after: J. Ehlers

River	Catchment	Area
Rhône	Geneva	8000 km²
Rhône	estuary	100500 km²
Mur	Graz	7080 km²
Danube	Vienna	101300 km²
Drava	Maribor	13300 km²
Salzach	Salzburg	4400 km²
Inn	Kufstein	9700 km²
Rhine	Basel	36200 km²
Rhine	Constance	10950 km²
Adige	Trento	9900 km²
Po	estuary	80000 km²

B. The Drainage System

Erosion by glaciers during the ice ages changed the morphology of the Alps significantly. Many rivers were diverted and drainage divides displaced. Only the rivers in those parts of the Alps that were never glaciated can be used to infer the drainage geometry in the Miocene. The few regions that remained ice free throughout this time are the Napf region of Switzerland (p. 242) and the Styrian massifs of *Hochschwab (p. 142)*, der *Koralpe* (p. 104) and part of the Seckauer Tauern where peaks are up to 2400 m high. For this reason, these massifs have rounded summit plateaus instead of the typical glacial cirques that characterise most Alpine peaks above 2000 m elevation. In all other regions of the Alps, glaciers carved the valleys so deep that it is near impossible to know if the valleys were wide or narrow or which direction the rivers flowed in them. Many pre-glacial river courses can merely be guesses: the river Inn may have flowed north through the Achensee, the river Salzach turned north in Zell am See and the lower courses of the Rhine and Rhône are completely unknown.

As a symbol for the recent geometry of the drainage system as we know it today, the photo at right shows the *Lunghin Pass* (2645 m) near the *Maloja lakes in the* Bündner Alps of eastern Switzerland. Here the catchments of the Po, Rhine and Danube meet. A drop of water that falls on the pass may end up in either the Adriatic Sea, the Black Sea or the Atlantic, depending on whether it flows into the Mera, the Inn or the Julia. The divide between Atlantic and Black Sea runs across the Julier Pass near Davos, through Piz Buin in the Silveretta to the Arlberg Pass and then through the Allgäu Alps. The drainage divide between Adriatic Sea (Po) and the Atlantic (Rhine) runs from the Lunghin Pass westwards to St. Gotthard Pass.

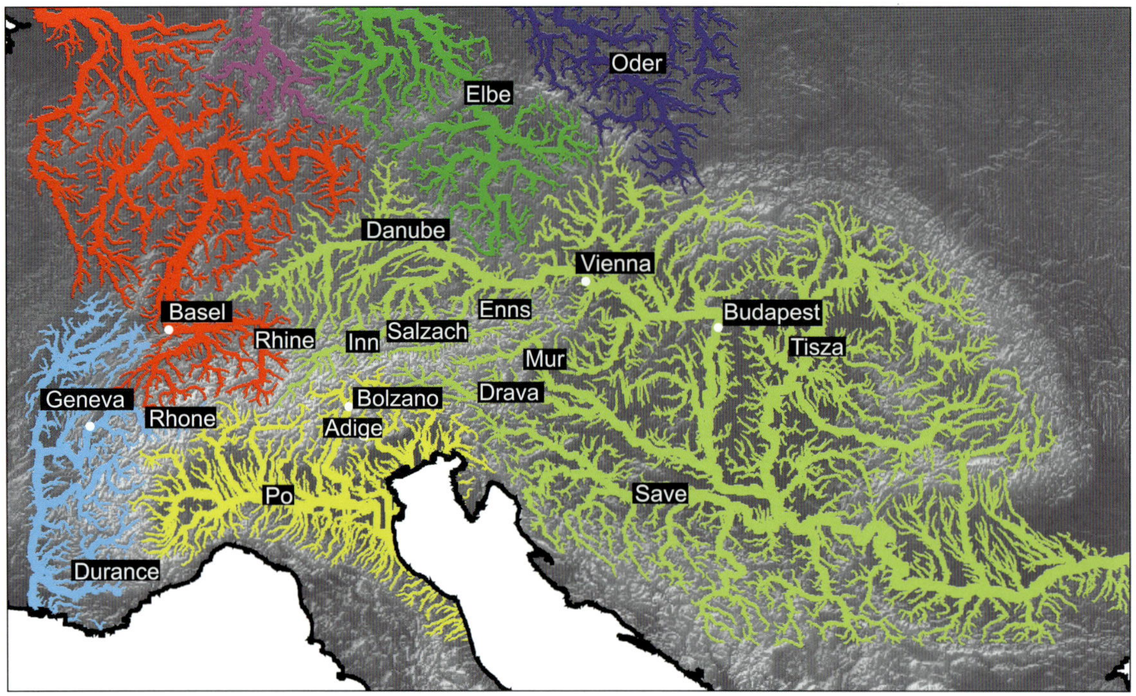

Above and at left: Catchments of some major rivers in the Alps. Red = Rhine, green = Elbe, blue = Oder, light blue = Rhône, yellow = Po, light green = Danube. The table shows the size of the catchments of some important rivers above a given point.

Page 117: The Maloja Lakes and the Lunghin Pass in the foreground. The pass forms the principal triple-divide of the Alps. The colours on the sketch show which areas around the pass drain into which catchments, using the same colour code as the map at left.

Ortler
Bernina
Inn
(Black Sea)
Gelgia / Julia
(North Sea)
Maloja Pass
Maira / Mera
(Adriatic Sea)

Above: The cable car up the granite needle of Aiguille du Midi (3842 m) is the second highest of its kind in Europe (with the highest going up the Kleine Matterhorn). It has been in operation ever since 1955 and is a popular departure point for climbers and scientists to explore the Mt. Blanc Massif.

The small photo shows the first hairpin curves of the road from the Rhône Valley into the Lötschen Valley. There, a 14 km long tunnel starts that connects the Rhône Valley with the Bernese Oberland.

Page 119: Map of the Alps showing 25289 earthquakes. Although this sounds like an unbelievably high number, the Alps are actually one of the seismically more quiet mountain ranges of this globe. The distribution of earthquakes is closely associated with the current motion of the Adriatic and European plates. Therefore, the highest density is in the region of Friuli. On the figure, the colour of the dots indicates the depth of the epicentres from near surface (red) to 110 km depth (yellow). The size of the dots indicates magnitude from 3 to 6.5 on the Richter scale.

C. Mankind

The Holocene covers the last 10000 years of the history of Earth: the time since the end of the last glaciation period. The glaciers retreated and life re-conquered the Alps. Man began to settle, use and study the mountains. The oldest mine in the Alps – a marble quarry near Vienna – started operating about 6000 years ago. During the bronze age, salt and copper mines were widely spread. Today, man has built his monuments up to the highest peaks of the range. The largest tunneling project in the world, the Gotthard Base Tunnel, is also in the Alps (p. 234).

Many of the major fault zones that shaped the Alps in the Miocene ceased to be active in the Holocene, and the valleys that had formed along them started being partially re-filled with sediments from the waning ice age: the *Rhône-Simplon Fault Zone*, the *Salzachtal-Ennstal Fault*, die *Inntal Fault* and the *Periadriatic Line*. However, overall, tectonism did not stop. The Alps are now being pushed as a whole over the top of the Molasse Basins in both the north and the south. Italy turns counterclockwise around the Italian city of Turin at a rate of 0.5° per million year. Because of this rotation, there is no convergence between Adriatic and European plates north of Turin. West of Turin the Alps are being pulled to the south and east of Turin the two plates converge. Going further east the relative motion between the plates increases and reaches its maximum in the Friuli region. The strongest historically documented earthquake of the Alps occurred in Basel in 1358. In the Eastern Alps, the strongest known earthquake occurred only 10 years previously near Dobratsch in the Austrian province Carinthia (p. 120).

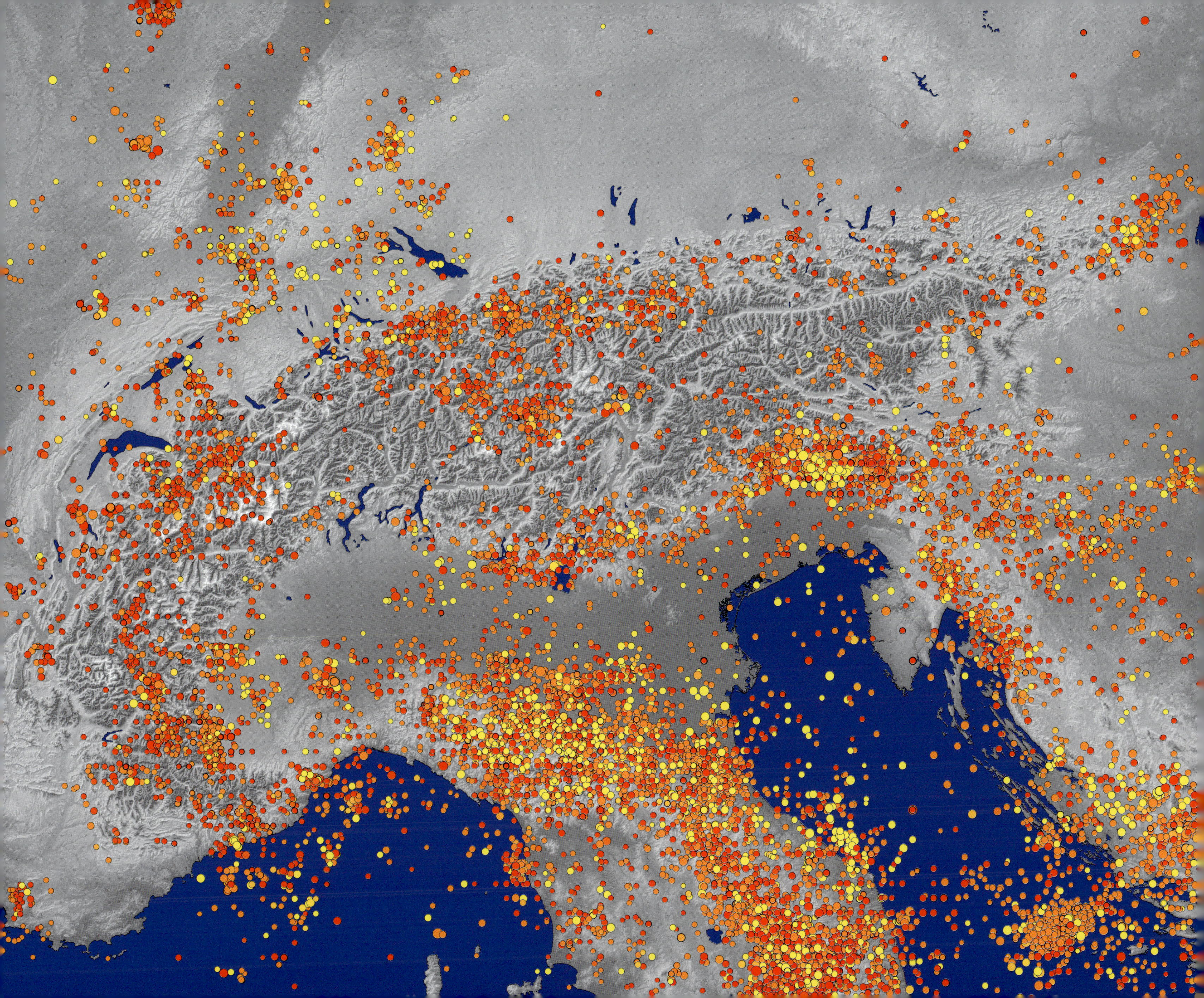

Looking west along the Periadriatic Fault (orange surface on sketch) over the Gail Valley onto Dobratch at right and the Southern Alps (at left). South of the river Gail there are late Variscan volcanics and sediments that belong to the Southern Alps. Behind them the Mesozoic limestones of the Julian Alps. After the last ice age the valley along the Periadriatic Line was filled with sediments, so that the valley floor is flat. One of the granites that intruded the fault zone is located in the background (red). At the very right the peaks of the Hohe Tauern range on the skyline (Großglockner, Hochalmspitze).

Dobratsch (2 166 m) is the local mountain of the Austrian city of Villach in Carinthia. It is also the location of one of the largest historic landslides in the Alps (p. 42). The slide was triggered by an earthquake in the Friuli region in 1348, and more than 150 million cubic metres of rock skidded from the summit region of Dobratsch into the valley below. The river Gail was dammed up and a three kilometre long lake formed that remained until the 18th century. Despite wide spread settlement of the Gail Valley at the time, the landslide itself did not bury any houses, as the mass slid onto the unsettled debris region of an even larger prehistoric landslide of almost 1 cubic kilometre of rock. Concerning the 1348 earthquake, there are reports that many churches collapsed in the Friuli region and that about 2000 – 5000 people died. Today the freeway that leads from Austria to Italy crosses the debris mass just west of Villach in the region of the little settlements of *Unterschütt* und *Oberschütt.*

Geologically, Dobratsch is part of the Drauzug, a range that is made up of Triassic limestone and Raibl beds similar to those of the Calcareous- and the Julian Alps. These limestones lie atop the Gailtal Basement which may be seen in many of the

surrounding massifs. Dobratsch Massif is surrounded on all sides by faults that have exposed it geologically and morphologically as a free-standing massif. Along the Gail Valley there is the Periadriatic Line, and another major fault zone runs along the Bleiberg Valley on the north side of the massif. The glacially carved valleys of Drau and Gail have carved along the soft rocks of the fault zone. At the end of the ice age, the valleys were refilled by sediments and are therefore flat today.

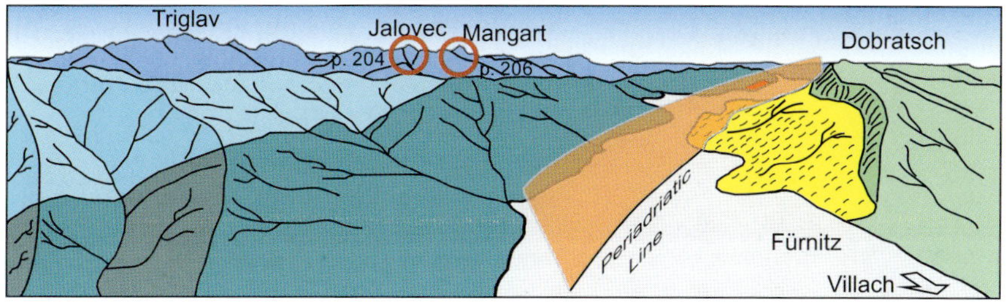

Right: The Lac de Dix dam on the northwest side of Grand Combin south of the Rhône Valley is 285 m high. It is the highest dam in the Alps and the fifth highest in the world.

Page 123: The Vajont dam – site of the largest man made disaster in the Alps – seen from the west. It may clearly be seen how the limestone layers on the right hand side of the valley are inclined parallel to the slope, causing the man-made landslide.

Below: The Großglockner Hochalpenstraße was opened in 1935, and was then considered one of the most ambitious construction feats in the world. The road crosses the entire Tauern Window and is a popular route for geological field trips. The photo below shows the Fuschertörl Pass (2450 m), Großglockner and Fuscherkarkopf in the background.

The influence of mankind on shaping the Alps is impressively symbolised by the largest man made catastrophe in the Alps: the landslide of Vajont, which occurred in connection with a hydroelectric project. The dam lies high above the township of Longarone at the western end of the Friuli Dolomites in the Piave Valley of northern Italy. It was supposed to create electricity from the small river of Vajont for the city of Venice. A superlative dam was planned and built in the fifties of the last century. The elegantly arced dam was 261 m high and could contain a volume of 152 million cubicmetres of water. With these extremes it was the highest of its kind in the world, until an even higher dam was built, also in the Alps: Lac de Dix.

Geologists had recognised early on that the inclination of the limestone layers parallel to the slope above the dam was a potential landslide hazard and engineers began careful partial fillings in 1960. Already during the first filling a landslide of some 700000 cubic metres of rock slid into the lake, but no damage was caused. A series of experimental partial fillings and drainings were done, but all of these were associated with mass movements on the surrounding slopes.

Despite all warning signs, the dam was eventually completely filled with water in 1963, and on October 9th more than 250 million cubic kilometres of rock thundered from the slopes of Monte Toc into the lake. The dam held, but 25 million cubic metres of water spilled over the wall and the flood wave killed about 2000 people in the village of Longarone below.

Today, the influence of mankind on re-shaping the Alps is mainly seen in a series of huge tunnelling projects, for example the Gotthard Base Tunnel or the Koralpe Base Tunnel.

bedding
Carbonates of the Southern Alps
Landslide mass
Slip surface

4. Geo-Highlights of the Alps

Some Geo-Highlights that are do not feature in this chapter have already been discussed in the context of chapter 1–3. On page 294–295 there is a complete picture reference of all images of the book.

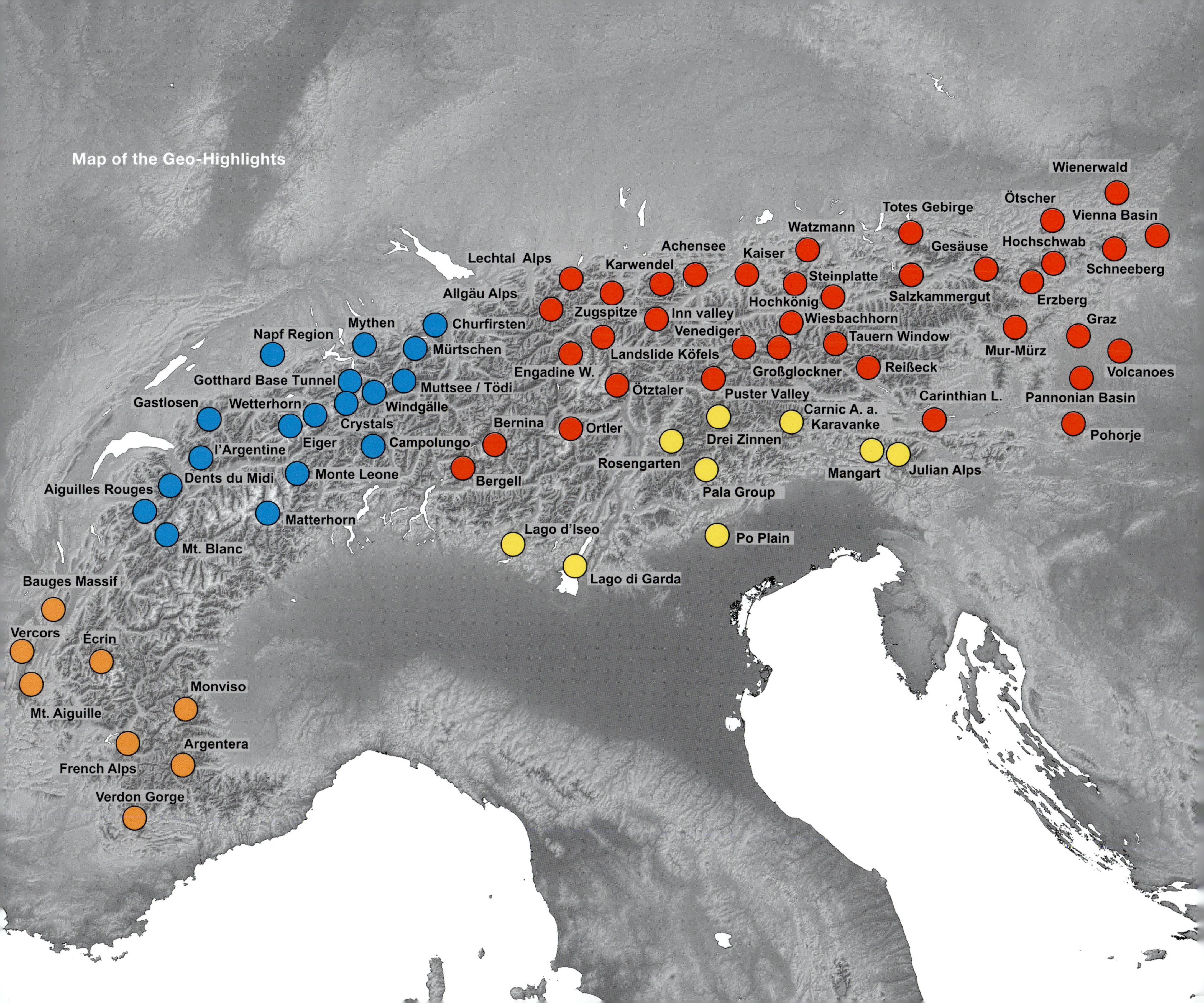
Map of the Geo-Highlights
Wienerwald
Ötscher
Vienna Basin
Totes Gebirge
Watzmann
Achensee
Kaiser
Gesäuse
Hochschwab
Schneeberg
Lechtal Alps
Karwendel
Steinplatte
Allgäu Alps
Hochkönig
Salzkammergut
Erzberg
Zugspitze
Inn valley
Wiesbachhorn
Graz
Venediger
Tauern Window
Mur-Mürz
Churfirsten
Mythen
Napf Region
Mürtschen
Landslide Köfels
Volcanoes
Engadine W.
Großglockner
Reißeck
Gotthard Base Tunnel
Muttsee / Tödi
Ötztaler
Carinthian L.
Pannonian Basin
Gastlosen
Wetterhorn
Windgälle
Puster Valley
Carnic A. a.
Karavanke
Crystals
Bernina
Ortler
Pohorje
Eiger
Campolungo
Drei Zinnen
l'Argentine
Rosengarten
Mangart
Julian Alps
Dents du Midi
Monte Leone
Bergell
Aiguilles Rouges
Pala Group
Matterhorn
Lago d'Iseo
Po Plain
Mt. Blanc
Lago di Garda
Bauges Massif
Vercors
Écrin
Monviso
Mt. Aiguille
Argentera
French Alps
Verdon Gorge

4.1. Eastern Alps

The Hohe Tauern are part of the Eastern Alps. View to west.

Right: geological map of the Pannonian Basin (red dot = photo above).

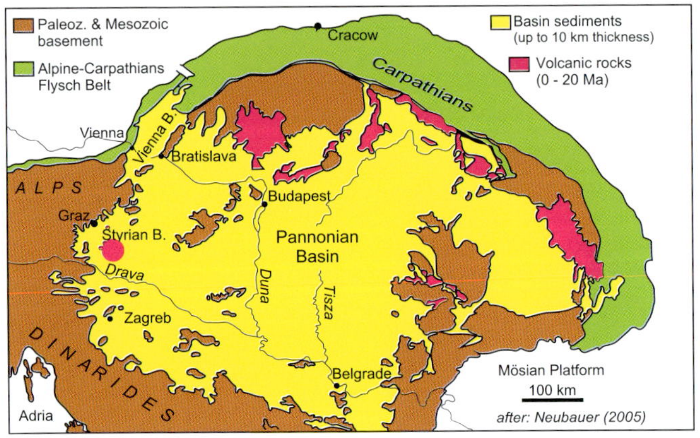

The Pannonian Basin: To the east, the Alps are bound by the Pannonian sedimentary basin. The westernmost lobes of this huge and almost round region are the Vienna and the Styrian Basins. The Pannonian Basin formed between the Alps and the Carpathians due to extensional thinning of the continent in the time between 30 and 10 Ma. The stretched lithosphere heated up and volcanoes penetrated the thinned crust. The Earth's surface slowly sank down. Salt water moved in and formed a sea called the *Paratethys*. Marine sediments began to accumulate, and the weight of these sediments pushed the bottom of this sea even further down. In the end, many

kilometre-thick sequences of sediments were deposited in the basin. About 7 million years ago the basin began to invert: the massively thinned crust stopped sinking down and began to lift up. The Pannonian sea became a fresh-water lake and finally dried up completely. The *Wildoner Berg* near Graz was pushed up. The basin sediments slowly eroded and the hilly landscape of western and eastern Styria formed. Finally, during the ice ages, rivers poured large gravel fans over the basin. At the transition from the Styrian hilly landscape to the Pannonian plains, the front ends of these gravel fans can be seen.

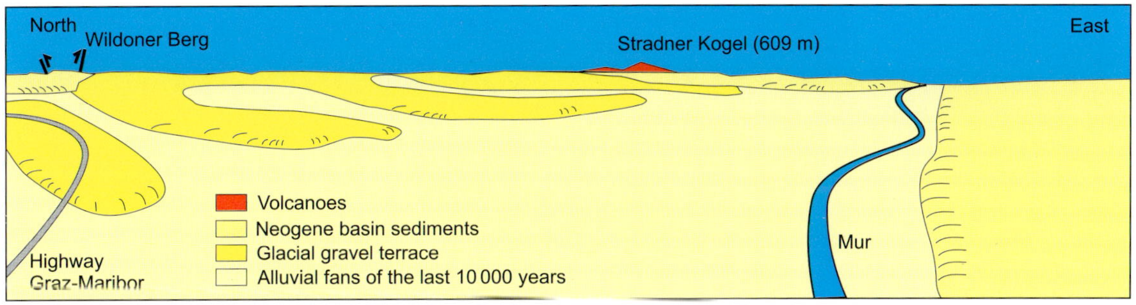

Geological sketch of the photo above.

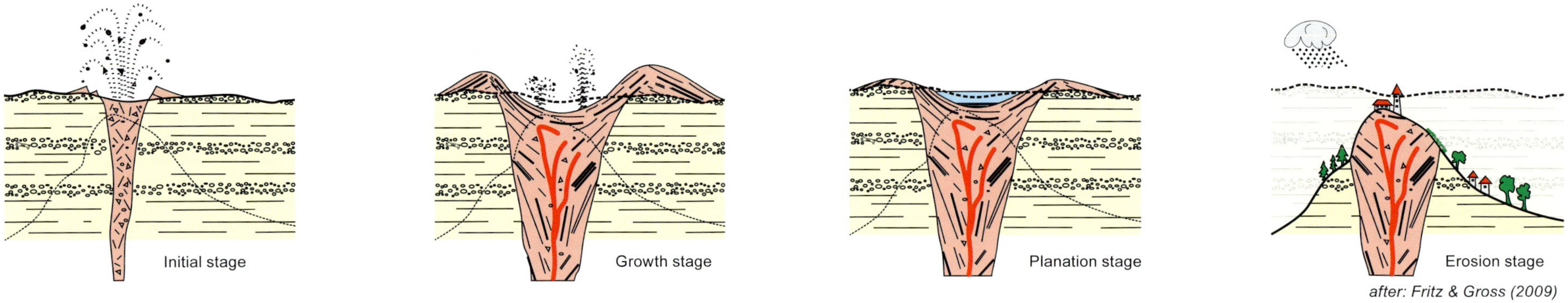

Above: Different stages in the development of volcanoes in the Austrian province of Styria.

Below: The rugged volcanic pipes of Eastern Styria were the perfect choice for a fortress in medieval times. The Riegersburg is the most famous of them all. It was built about 850 years ago on the volcanic tuff of the Riegersburg volcano on top of a stone-age settlement. "Olivine-bombs" are not really a secret defense weapon, but are a popular collectors item in the region and testify to the deep-seated nature of the magma.

Volcanoes along the Eastern Margin of the Alps: The Austrian province of Styria (Steiermark) is not only a well-known wine growing region, but also one of the few volcanic provinces of the Alps. There are two phases of volcanism, both of which occurred in connection with the development of the Pannonian Basin east of the Alps. The older phase produced a calc-alkaline basaltic lava around 14 – 16 million years ago. The chemical composition of this lava indicates that it formed in connection with a subduction zone or with extension of a tectonic plate.

These older volcanoes are of the same age as the marine sediments that surround them, so we must imagine the Styrian Basin at this time as a tropical sea with islands of smoking volcanoes. The volcanoes in *Gleichenberg* and the volcano near *Weitendorf near Wildon* belong to this group of volcanoes.

The younger generation of volcanoes is not even 2 million years old. Geologically speaking this is practically still active, and many hot springs in the Styrian region still testify to high heat flow in the region. This younger generation of volcanism produced a very different type of lava: a lava that formed deep in the Earth's mantle. During its ascent through the lithosphere, this lava tore mantle fragments from the wall with it. These fragments may be found as "olivine bombs" in the Eastern Styria region. The *Riegersburg* and *Kapfenstein* volcanoes belong to this group.

Despite their young age, not much of the shape of these volcanoes is preserved today. This is because the Styrian Basin stopped sinking down about 7 – 10 million years ago. The basin began to lift up at several tenths of a millimetre per year. Slowly it is being incorporated into the mountain building of the Alps. During the uplift, layers of sediments about 1 kilometre thick were eroded from the region and the hilly landscape of the Styrian wine growing region formed. The hard tuffs and lavas of the volcanic pipes were excavated and now form the highest points of the landscape.

Page 131: The Kapfenstein volcano in the Eastern Styria region in the southeast corner of Austria. This volcano belongs to the younger generation of volcanoes in Styria. The winery Winkler-Hermaden on its slopes produces a wine called "olivin", named after the olivine nodules found in the region. The winery is well known among geologists, also for the fact that it is run by the family of one of Styria's most famous geologists: Arthur Winkler von Hermaden was a well known geologist in the middle of the last century.

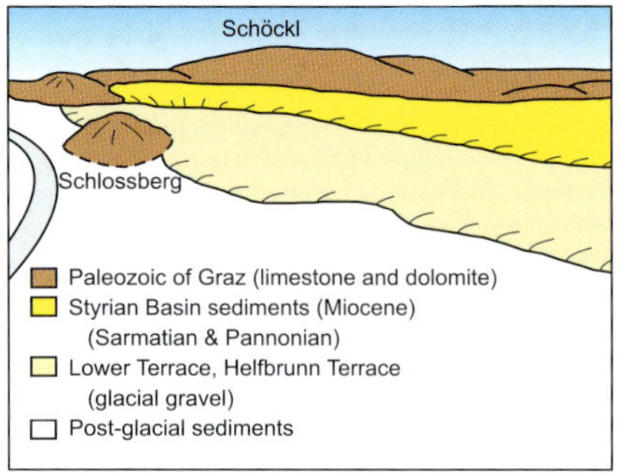

Sketch above and page 133: The centre of Graz, capital of the Austrian province Styria. The Schlossberg in the city centre and the hills in the background (with the local mountain of Graz: Schöckl) are both made up of carbonates that belong to the Paleozoic of Graz. The Schlossberg is surrounded by Neogene basin sediments and river terraces from the last ice age.

Right and top right: Geological map of the Styrian Basin. The geological profile is schematic and corresponds to a cross section through the photo on p. 133.

Graz – City Between Mountains and Basin: The capital of the Austrian province of Styria has a special geological position: it lies on the boundary between the Alps and Europe's most famous sedimentary basin, the Pannonian Basin. The hills to the north and west of the city (e.g. *Plabutsch, Schöckl*) are part of the Alpine basement made up of ancient carbonates that were metamorphosed and deformed as part of mountain building processes of hundreds of millions of years. In contrast, the hills in the east are made from marine sediments deposited into the *Paratethys Sea* that filled the Pannonian Basin in the Neogene. The westernmost part of this basin – the Styrian Basin – was filled by the Paratethys sea until less than 12 million years ago. One of the youngest marine deposits that record this is the *Leitha limestone,* a coral reef limestone deposited along the margins of this sea 15 million years ago. Later, the basin inverted and the surface lifted from the

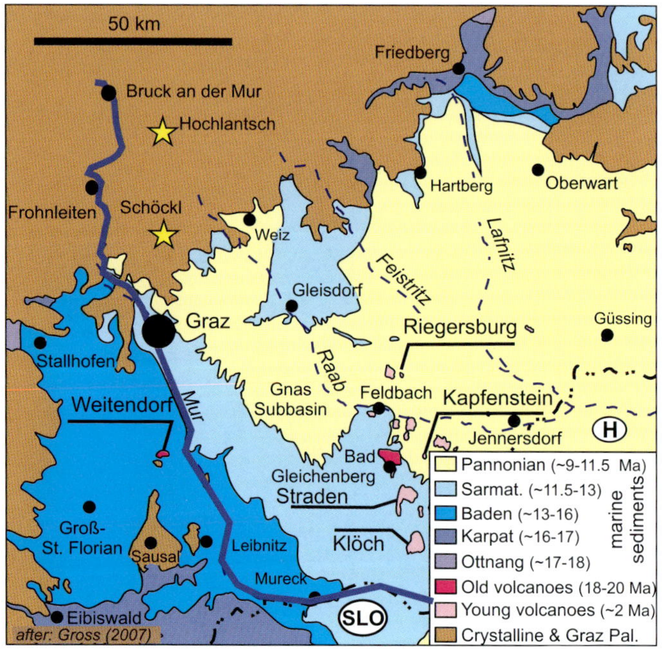

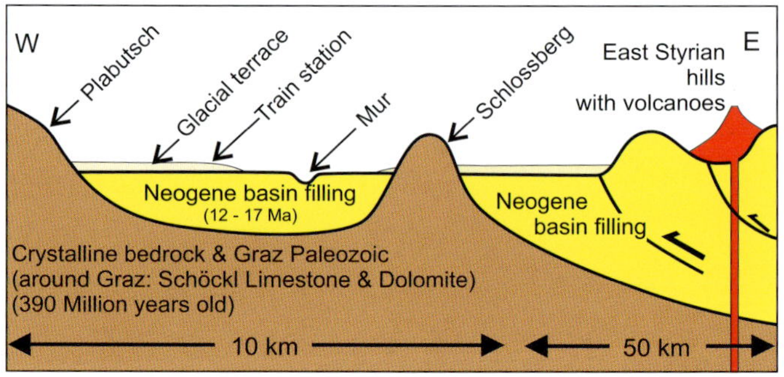

sea. The uplift of the basin in the last 5 Ma is responsible for the fact that the hills on the west and east side of Graz are of completely different geological origin. Hills like the Schlossberg in the centre of the city are like islands of the basement that penetrate the Neogene sediments. The Alpine basement just north of Graz belongs to the so called *Paleozoic of Graz*. This is a tectonically isolated slab of rocks a few kilometres thick that is made up of Paleozoic, low grade carbonates and schists that sit on top of much younger garnet mica schists of the Austro-Alpine crystalline basement of the Kor- and Gleinalpe. The rocks of the *Schöckl-* and *Hochlantsch Massif* are part of this *Paleozoic of Graz*. Similar rocks occur in the Alps only in a few regions (see p. 94 for listing).

The region around Graz also provides insights into the most recent part of the Alps' geological history. Kilometre-long cave systems (e.g. *Lurgrotte* or *Drachenhöhle*) formed in connection with the incision of the valley of the river Mur. Many of the caves contain quartz pebbles that were washed in during their formation. Geologists have dated this event and reconstructed the landscape evolution of the region.

Caves, a mild climate and the absence of glaciers during the ice ages are also reasons why the oldest archaeological finds in the Alps were made in the Graz region.

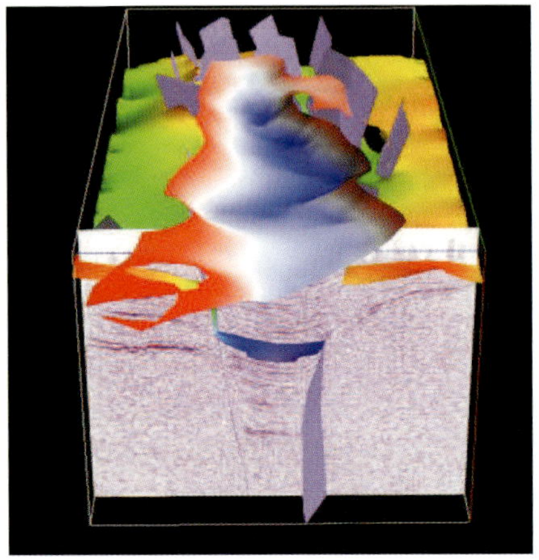

Above: Block diagram of the Vienna Basin (after K. Decker and R. Hinsch). The sedimentary layering and faults are well resolved by seismic studies.

Vienna Basin: The Vienna Basin is a deep sedimentary basin at the northeastern corner of the Alps. It is about 200 km long and 50 km wide. Its shape is typical for "pull apart basins" (see p. 109) as they form along strike slip zones. The so called *Vienna Basin Margin Fault* runs along the west side of the basin along the *Wienerwald*. The fault scarp may be well seen from the freeway leaving Vienna to the south. In the east the fault zone that bounds the basin runs along the western margin of the *Leitha Range*. 18 million years ago, movements along these two fault zones opened up the Vienna Basin and the *Paratethys sea* flooded in. However, just as rapidly as the basin floor sank, the basin was filled with sediments, and the sea of the Paratethys remained shallow. Thus, the seven kilometre thick sediment pile is entirely made up of shallow water sediments. At about 11 Ma the basin inverted: the surface began to rise and the basin was disconnected from the Paratethys sea. The remaining lake, *Lake Pannon*, bagan to turn to fresh water and finally disappeared completely about 9 million years ago. The youngest Miocene basin sediments are huge delta fans of the Danube river which had its bed some 50 km further north at that time. Then all tectonic activity came to a halt. The Danube was diverted to its present day position and formed large river terraces in the central part of the Vienna Basin.

About 500 000 years ago the fault zones that opened the basin in the Miocene became active one last time. The basin floor started sinking again and river sediments of the Danube were deposited up to 200 m thick. GPS-measurements show that the sides of the basin still move at about two millimetres per year with respect to each other. This recent phase of tectonic and sedimentary activity is responsible for the flat landscape in the basin region. The geology of the Vienna Basin is very well known because of its economic importance: the basin contains some of the largest Central European oil and gas reserves.

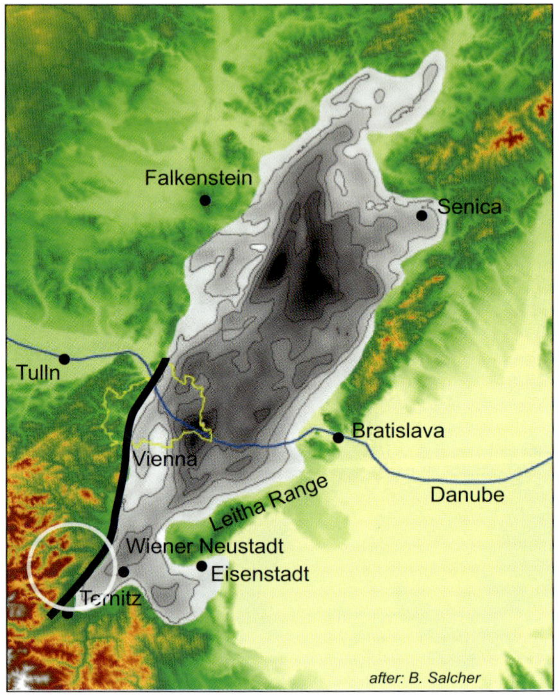

Above: The position of the Vienna Basin superimposed onto a topographic map of the eastern end of the Alps. The gray shading indicates the thickness of the sedimentary strata between 100 metres (white) and 6 000 metres (black). The thick black line is the Vienna Basin Margin Fault. Other fault zones delimit the basin along its eastern margin (ring = photo p. 135).

Page 135: Flying around the north corner of the Schneeberg (2 076 m) the Vienna Basin emerges on the horizon. The basin margin is clearly seen as the topographic step from the flat basin floor to its hilly surroundings. The township in the image centre is Puchberg am Schneeberg. To the left and behind it Wiener Neustadt can be seen in the middle of the Vienna Basin.

The northern part of the Pohorje Massif (Kozjak Range) looking east towards Maribor in Slovenia. The highest peak of the massif (Črni Vrh, 1543 m) can be seen in the background along the right hand photo margin. In the left hand background is the river Mur and the Southern Styrian Basin of Austria. The steep valley of the river Drava in the foreground was only carved recently. Only a few million years ago the river still flowed in the Ribnica Trough, the valley in the right half of the image.

The Pohorje Massif in Slovenia: This massif is the southeasternmost range of the Alps and lies directly on the plate tectonic boundary between European and Adriatic plates: the Periadriatic Line. To the north (at the Radl Pass) and east (near Maribor) the massif disappears underneath the sediments of the *Pannonian Basin* and to the west it is bound by the *Lavanttal Fault* (p. 108). Between Dravograd and Maribor the massif forms a huge elliptically shaped dome. In the Cretaceous (at about 90 Ma), most of the massif was still buried about 80 – 100 km deep in the crust where it was almost 700° C hot. It was then that the rocks were metamorphosed to garnet-bearing mica schists and gneisses. In the southeastern corner of the massif high pressure minerals have been found that formed in up to 100 km depth (see also p. 104, 272). After that, the massif cooled and slowly rose to the surface. 19 million years ago a granite body intruded. This granite is the easternmost of the *Periadriatic intrusives* (p. 110) but it is substantially younger than all the others (which intruded

about 30 Ma). Because of its position on a plate boundary, the Pohorje Massif is tectonically still active. Geomorphological features have been used by geologists to show that the massif only uplifted in the last few million years. For example, the Drava river runs from the broad Lavant Valley into the steep Drava gorge, indicating that the uplift of the Pohorje Massif may be younger than the course of the river.

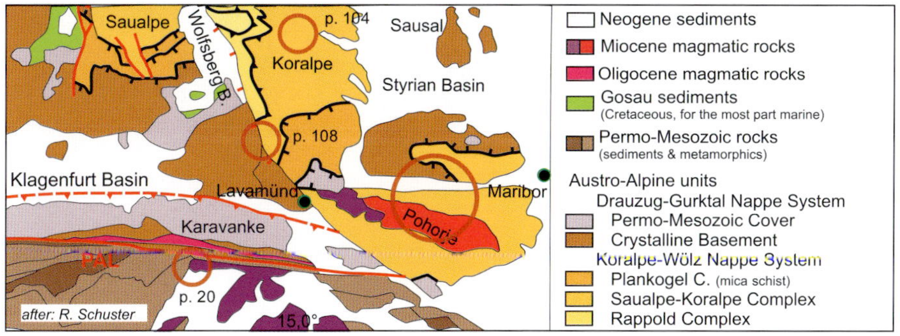

Right: Geological map of the eastern margin of the Alps. The large red ring shows the position of the photo.

Above: Ötscher (1 893 m) is one of the easternmost mountains of the Alps. The settlement Pfaffenschlag in the foreground. The township of Lunz am See is just outside the image to the right.

Right: Geological cross section through the Calcareous Alps (purple), Flysch Zone (green) and Molasse Zone (yellow).

Page 139: The easternmost part of the Kalkvoralpen near the Flysch Zone. The inset shows a geological cross section through the northern Molasse Zone, the Flysch Zone and the Northern Calcareous Alps (roughly in the region of the Tote Gebirge).

Alpenvorland – Wienerwald and Ötscher: The northeastern boundary of the Alps is the *Ötscherland* region of Lower Austria and the *Wienerwald* region. Ötscher is 1 893 m high and is one of the easternmost and at the same time northernmost mountains of the Alps that has anything like an Alpine character. Geologically this part of the Alps belongs to the Northern Calcareous Alps that are called *Kalkvoralpen* in the region north of Mariazell, Schneeberg and the Vienna Basin. At their northern end, the Kalkvoralpen are being thrust over the top of the Flysch Zone on the level of Waidhofen an der Ybbs, Scheibbs and Wienerwald. During the Cretaceous and the Paleogene, the Kalkvoralpen were internally hacked up by faults. Today, they lie in a series of nappe-like lobes stacked up from south to north over the top of each other (see map below). The typical sequence of rock types known from elsewhere in the Calcareous Alps remains preserved in each of these lobes: at the base there is usually *Werfen Formation*. Above this there is Triassic Hauptdolomit, Dachstein limestone, Jurassic marls and Muschelkalk limestone.

The typical landforms of the region stem from the fact that the region was never glaciated. As such, there is no glacial carving of valleys. Steep gullies instead of broad valleys are typical. A large part of the region was covered during the ice ages by thick loess deposits (air blown dust).

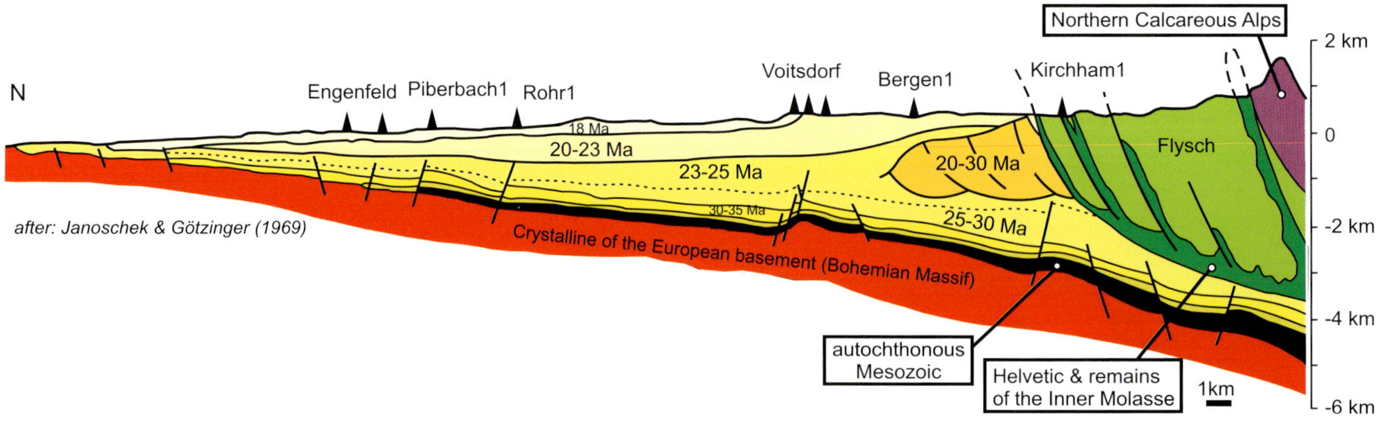

N
S
m
1000
500
Haigerkgl. (884 m)
Weherberghof
Falkenstein-E
Schindelhütte, Erlauf (436 m)
Beykogelhof
Schönwald
Firmböndl (1090 m)
Ranegg
Brandkogel (1094 m)
Liasfleckenmergel
Ötscherkamm-W (1599 m)
Wetterstein limestone
Opponitzer Rauhwacke
limestone
Hauptdolomit
Lunz beds
Räth
Radiolarites
Muschelkalk (Anisian)
?
Werfen beds
Haselgebirge (Upper Perm.)
?
after: Tollmann (1965)
Dachstein limestone
Hauptdolomit (Nor.)
2 km

Left: The north faces of Rosskuppe (2 152 m) and Dachl include some of the most famous climbing routes in the Gesäuse mountains.

Right: Geological map of the eastern sector of the Northern Calcareous Alps.

Page 141: The north faces of the Gesäuse Massif seen from the west.

The Gesäuse Group: The Gesäuse Massif is a group of rugged limestone peaks on both sides of the Enns Valley between the settlements of Admont and Hieflau in the Austrian province of Styria. The massif includes the *Buchstein Group* (2 224 m), *Hochtor Group* (2 369 m), *Admonter Reichenstein Group* (2 251 m) and the *Lugauer-Zinödl Group* (2 217 m). Although none of these mountains is very high in comparison to the peaks of the Central Alps, their north faces drop more than 1 000 m vertically into the Enns Valley and are among the highest of their kind in the Eastern Alps. Geologically, the Gesäuse mountains are part of the *Mürzalpen Nappe* of the Northern Calcareous Alps. This nappe also includes the Hochschwab Massif (p. 142), Tonion, Schneealm and the Hohe Wand in Lower Austria. On top of the Mürzalpen Nappe the Schneeberg Nappe lies like an island. This upper nappe makes up the mountains *Rax* and *Schneeberg* (p. 135). In the west, the Mürzalpen Nappe lies near Admont on top of the *Warscheneck Nappe*.

At their base, the Gesäuse mountains are made up of *Wettersteinkalk* (Middle Triassic: 235 Ma – up to about 1 200 m elevation). Above this, the schist band of the *Raibl Formation* (220 Ma) follows. Above that there are several hundreds of metres of Hauptdolomit (215 Ma). The uppermost 500 vertical metres are made up of the thick banked solid Dachstein limestone (220 Ma) that is so popular among climbers. All rocks were deposited in a shallow sea and contain a corresponding fossil content. In particular the *Kuhtrittmuschel* and fossil algae are common findings along the hiking trails that cross the Dachstein limestone.

The Gesäuse mountains also have a fascinating recent geological history. Since about 20–30 Ma a huge fault zone has been active along the Enns Valley (p. 150), and many of the rocks in the lower reaches of the mountains are ground up to a mosaic-like brittle limestone mass. During the ice ages a huge block of ice dammed up the entrance of the Gesäuse gorge. The Enns was re-directed across the Buchauer Saddle towards St. Gallen and a lake formed that reached all the way back to Liezen. Lake sediments still cover the region of the former lake. The Gesäuse mountains are also known for their pre-historic copper mining.

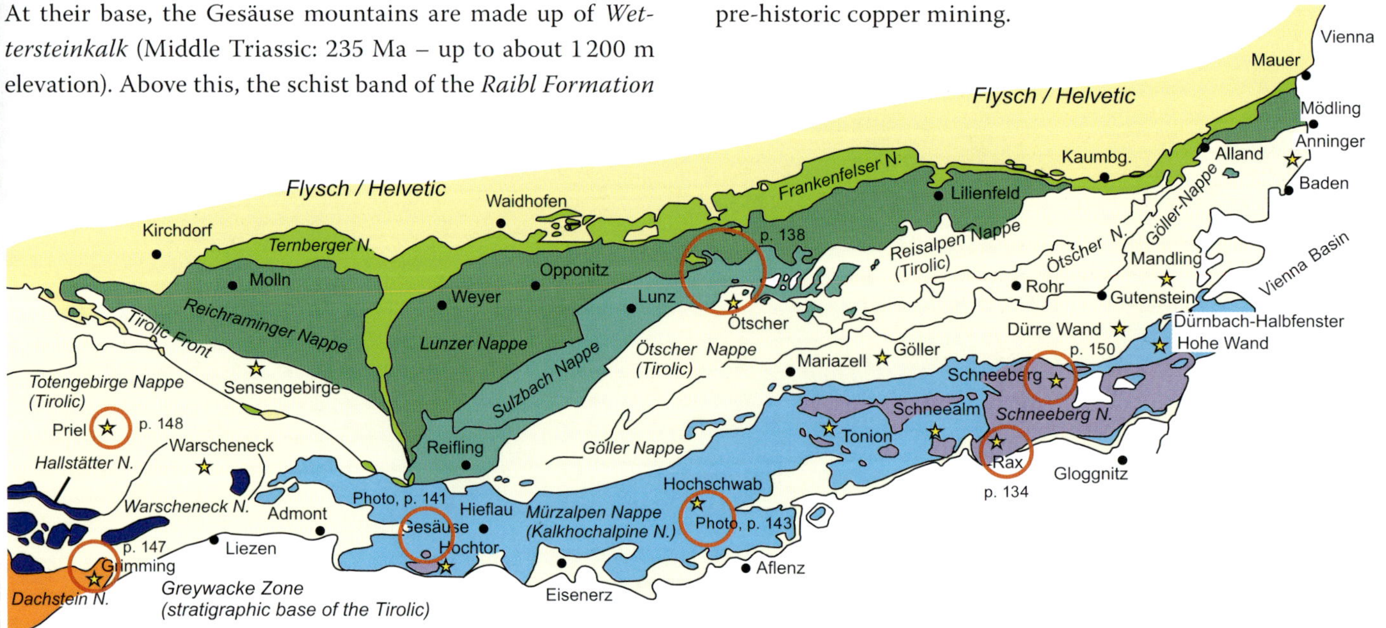

after: Tollmann (1965)

Planspitze (2117 m)
Dachl
Hochtor (2369 m)
Ödstein (2335 m)
Reichenstein (2251 m)
Sparafeld
Gesäuse
Johnsbach Valley
Dachstein limestone
Hauptdolomit
Raibl / Lunz beds
Wetterstein limestone / Ramsau dolomite

Right: The entrance to the Melkboden ice cave not far from the Trawies Pass in the central Hochschwab Massif. Just behind the visible edge in the cave a 540 m vertical drop starts that is about 20 m in diameter. The shaft is one of the deepest free abseil stretches in Styria (poto: S. Oswald).

Page 143: The south side of the Hochschab Massif and the Trawies Valley seen from the west. The lowest most limestone face on the photo is the Hundswand. In front of it lies the geological half window of the Bodenbauer region where the water impermeable Werfen schists, that form the base of the Hochschwab Massif reach the surface. The peaks and gullies from left to right are: Schönbergkar, Beilstein (2015 m, difficult to see as its not against the sky), Rauchtal (in the shade), Stangenwand (2157 m) and Zagelkogel (2255 m). The summit of the Hochschwab itself (2277 m) is the flat topped peak in the background behind the Trawies Pass. Karlhochkogel (2096 m) is the characteristic peak at right and the Festlbeilstein (1847 m) near the right hand photo margin. The entrance to the Melkboden ice cave lies to the right of the Trawies Pass.

The Hochschwab Massif in *Upper Styria* is also part of the Northern Calcareous Alps. The massif connects the Semmering Pass (984 m) in the east with the Präbichl (1226 m) in the west and forms the principal drainage divide of the Alps in this region. Because of its wide open planes above 2000 m elevation the massif is known for sudden weather changes and many hikers have become lost in the range. Tectonically, the Hochschwab Massif is part of the Mürzalpen Nappe. Akin to many other parts of the Calcareous Alps, the mighty limestone slabs of the massif lie on top of the schists of the *Werfen Formation* of 245 Ma age. The Werfen beds are a characteristic rock type that marks the beginning of the shallow water sedimentation in the entire Alps that then continued throughout the Mesozoic. Above the Werfen beds there is *Gutensteiner dolomite* several hundreds metres thick and above this there follows a slab of *Wetterstein limestone* several hundreds metres thick, that was deposited in the middle Triassic (230 Ma). This Wetterstein limestone makes up the Hochschwab south face. The highest parts of the massif, for example the summit region of the Stangenwand, is made up of the thickly banked *Dachstein limestone.*

The limestones of the massif are strongly karstified and many caves are developed. More than 3000 registered caves are known in Styria, more than in any other Austrian province, and a good part of those are in the Hochschwab Massif. One of the largest cave systems of the massif is the *Frauenmauer-Langstein* system that measures more than 20 km. However, most cave systems are as yet unexplored. The *Melkboden ice cave* is the deepest vertical shaft in Styria. It is 540 m in vertical extent. Only 50 years ago this cave would have been among the deepest known caves in the world. However, despite the huge entrance of the cave, the shaft was only explored in 1998. Because of the karstification, most water that rains on the massif drains to lower levels inside the mountain through

the limestones and there are few springs on the high planes. The water accumulates at the water impermeable schists of the Werfen Formation below the massif. In places where the Werfen beds reach the surface on the sides of the massif, there are large karst springs and a series of lakes: Grüner See, the swampy meadows around Bodenbauer, Leopoldsteiner See. The dramatic fluctuations of the water level in these lakes is in part due to the rapid turnover of the water budget in the karst massif and partly due to the usage of the water by humans. 100 years ago Emperor Franz Josef built one of the first major water pipelines of the Alps from the Hochschwab Massif to Vienna to supply the capital with water. Even today, the north side of the Hochschwab massif is one of the main water suppliers for the Austrian capital.

Page 145: The Styrian ore mountain (Erzberg) from the north. The township of Eisenerz (meaning: iron ore) is in the foreground. The peak in the clouds is the Eisenerzer Reichenstein (2 165 m). In the left third of the photo lies the Präbichl Pass (1 226 m), the principal water divide of the Alps in this region. The open pit mine operates on 25 levels of 24 m height each. The upper levels are higher than the Präbichl. Thus, in historic times, ore from the upper levels was brought across the pass to the south to the smelters in Leoben / Donawitz, while ore from the lower levels was brought northwards to the steel plants near Linz.

Above right: The Frauenmauer (1 827 m) is part of the Hochschwab Group. Behind it the Brandstein (2 003 m). Both mountains are part of the northern Calcareous Alps opposite the Erzberg. At the base of the Frauenmauer the Calcareous Alps lie on the Greywacke Zone.

The Styrian Erzberg: This iron mine in the Austrian province of Styria is the largest open cut mine in central Europe and the largest siderite deposit in the world. It is also possibly one of the longest continuously operated mines in the world. Since the beginning of the iron age more than 2 000 years ago, iron ore has been excavated from the site almost every day. Despite various crises and a low iron content of the ore, more than one million tons of ore are still being mined every year from the mountain and more than 100 million tons of ore are still present.

The ore mountain lies between *Hochschwab*, *Eisenerzer Alpen* and the *Gesäuse* mountains near the Präbichl – a pass that forms the principal divide of the Alps in this region. Geologically, the Erzberg is made up of a sequence of volcanic rocks and carbonates that are part of the Greywacke Zone of the Austro-Alpine. Below the Erzberg lies a fine-grained, reddish volcanic rock that erupted almost 500 million years ago. Above this, various limestones (called Rohwand) are interlayered with schists (300 – 350 Ma of age). These Paleozoic limestones bear the ore. The structurally highest unit are the purple-coloured Präbichl layers that are equivalent to the Werfen beds. The juxtaposition of Mesozoic shallow marine carbonates of the Northern Calcareous Alps and the Paleozoic carbonates of the Greywacke Zone is spectacularly exposed on the mountain opposite the Erzberg: the Polster (1 911 m). All the surrounding peaks, like the Frauenmauer or the Pfaffenstein, are part of the Northern Calcareous Alps.

The ore mined on the mountain is siderite ($FeCO_3$) and ankerite (Fe, Mg, Ca) CO_3. Even in their purest forms, these minerals contain substantially less iron than the minerals hematite (Fe_2O_3) or magnetite (Fe_3O_4). In most of the major iron mines around the world these latter two minerals are being mined, and the Styrian ore mountain therefore has a hard time keeping up. The siderite of the mountain was formed from the Paleozoic limestones of the Greywacke Zone. Iron rich fluids that probably came from the volcanic rocks below circulated through the limestones and already transformed the $CaCO_3$ to $FeCO_3$ (siderite) 300 Ma ago.

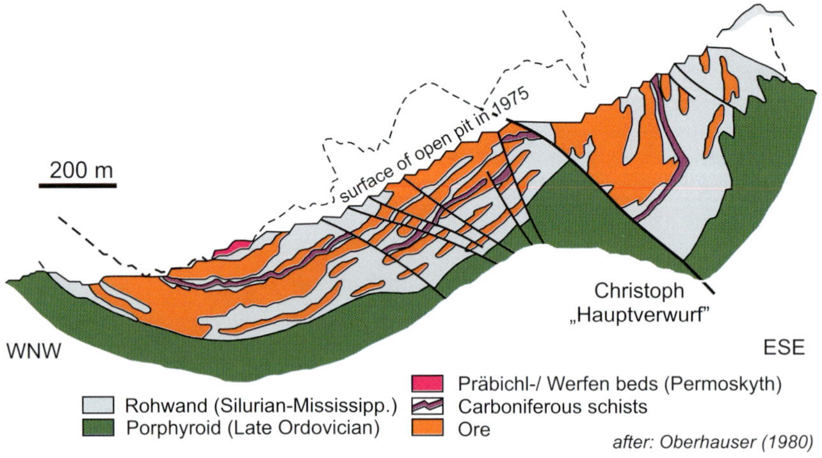

Below at right: Geological cross section through the Erzberg.

Above: The Dachstein Massif from the north.

Below: The Grundlsee lake and the mystic Toplitzsee (behind) seen from the west. Trisselwand (1754 m) and Backenstein (1772 m) form the cliffs on the left photo margin. They are made from Jurassic limestones. Behind the Toplitzsee lies the karst plateau of the Tote Gebirge with the summits Ofenkogel (2001 m), Hochkasten (2352 m) and Weiße Wand (2198 m). This entire plateau is made up of Triassic Dachstein limestone. The boundary between the Triassic and Cretaceous limestones lies on the shoulder several hundreds of metres above the Toplitzsee.

The Salzkammergut in the northwestern most part of the Austrian province of Styria is one of the most beautiful landscapes of the Eastern Alps. About a dozen lakes lie in a spectacular setting between rugged, karstified massifs of Triassic limestone: the *Tennengebirge*, *Dachstein* and *Tote Gebirge*. Geologically, the Salzkammergut region is part of the Northern Calcareous Alps, but nowhere else in this range the different Mesozoic rocks are in such close spatial proximity. Salt mines have operated here since the stone age and give the region its name. These salt deposits mark the beginning of the geological evolution of this part of the Alps. They formed at the end of the Permian, about 270 Ma ago, in salty lagoons that formed on the sinking continent.

When the continent sank further down, carbonate platforms grew on top of the salt deposits, similar to those growing in the Bahamas today. In the Jurassic (200 Ma), the ocean floor sank more rapidly and deep sea sediments (radiolarites, deep-sea limestone and schist) were deposited over the top of the drowned coral reefs. The Triassic carbonate platforms are preserved in the Dachstein and Tote Gebirge. The Jurassic deep water rocks today form the summit of Loser and island-like outcroppings on the plateau of the Tote Gebirge and in the valley of Bad Mitterndorf. In the Cretaceous, the sea became more shallow again and shallow water carbonates like coral reefs were deposited again. During Alpine deformation (about 160 – 90 Ma), the entire sediment stack was folded so that the old Triassic limestones were partly pushed over the top of the much younger radiolarites (for example near Grimming). Salt layers served as a gliding surface where folding is the strongest. This may be seen in some of the salt mines in the region of Hallstatt. In other places, for example on the Loser, the sediment stack of radiolarites and limestones was barely deformed, and it is still possible to walk through the entire Mesozoic sedimentary succession from the bottom to top.

Right: The valley of Bad Mitterndorfer looking south. The wide valley in the background is the Enns Valley. The mountains with snow in the background are part of the Upper (formerly Middle) Austro-Alpine of the Rottenmanner Tauern. The massive massif of the Grimming (2351 m) dominates the entrance of the Salzkammergut and is made up of Dachstein limestone. Its vertical banks appear to roll from the right background towards the front like a giant steam roller. In the process the Triassic limestone is pushed over the much younger Jurassic radiolarites in the foreground. In the centre of the photo the ski-jump of the Kulm can be seen, where several world championships have been held. On the map at right the radiolarites are mapped as "Hallstatt Zone".

10 km
Gmunden
Traunsee
Flysch Zone
Mondsee
Attersee
Photo, p. 148
Höllengebirge
Almsee
Wolfgangsee
Totes Gebirge
Altaussee
Grundlsee
Toplitzsee
Hallstattsee
Mitterndorf
Photo, p. 147
Dachstein
Grimming
Greywacke Zone
Ramsau
Gröbming
Hallstatt Zone (Jurassic)
Gosau Basin (Cretaceous)
Cenoman „Randschuppe"
after: Tollmann (1976)
Dachstein Nappe (Trias.)
Warschenegg N. (Trias.)
Tote Gebirgs N. (Triassic)
Reichraming N. (Trias.)

Traunsee and Tote Gebirge Range: To the north, the Salzkammergut region is bound by the mountains of the Tote Gebirge and the lakes *Traunsee, Attersee, Mondsee* and *Irrsee*. The Tote Gebirge mountains are made up of Triassic carbonates. However, the lakes and the shape of the Traunstein peak (1 691 m) give evidence for a much younger part of the geological history: Neogene tectonics and glacial carving. Most lakes in the region lie on fault zones in small tectonic basins that were initiated between 25 and 15 Ma ago in connection with the eastwards extrusion of the Alps. At this time the Northern Calcareous Alps were hacked up into individual blocks by many fault zones. These blocks today form the individual mountain massifs. This tectonic phase is also responsible for the beginning surface uplift in many of the massifs and the beginning of overthrusting of the Northern Calcareous Alps over the Flysch Zone. The mystical *"Augensteine"* – rounded quartz pebbles that are found on the summit plateaus of the Tote Gebirge and other limestone massifs in the region – show that rivers once flowed over these plateaus.

During the glaciation periods, the valleys were carved deeper. It is for this reason that the shape of the lakes resembles that of the large north Italian lakes (p. 212): long, north-south oriented lakes of which one end lies deep inside the mountain range and the other reaches into the plains of the Alpine foreland. The glacier that filled the Traunsee until about 10 000 years ago once reached to about 10 km north of the lake as far as Laakirchen.

Above: The main peaks of the Tote Gebirge seen from the south. From left to right these are: Schermberg (2 396 m), Großer Priel (2 515 m) and Spitzmauer (2 446 m).

Below and page 149: The Traunsee. On page 149 the view is looking south. The township of Gmunden (with the castle Orth in the lake) lies in the foreground. On the left background the Tote Gebirge. The Traunstein (1 691 m) right by the lake is made up of Triassic Wettersteinkalk. Tectonically, the Traunstein forms the northern end of the Höllengebirgs Nappe. It may clearly be seen that the limestone banks on the Traunstein dip towards the south and lie on top of the Grünberg (the forested peak with the radiotower) in the north (front). Grünberg is part of the Flysch Zone of the Alps.

The photo on this page is looking north. The township of Ebensee is in the foreground. In the far background, the Molasse Zone.

Schneeberg (2076 m) is one of the easternmost peaks of the Alps that exceeds 2000 m in elevation. Geologically it is part of the Calcareous Alps and sits right on top of a huge fault zone. For this reason the Mesozoic limestones of the massif are very fractured. The photo shows the mountain from the southwest where it rises high above the Höllental Valley in the foreground. In the background at left the hills of the Wienerwald. Just to the right of the summit the Vienna Basin in the background. The red dot on the map shows the position of the photo (G. Z. = Greywacke Zone).

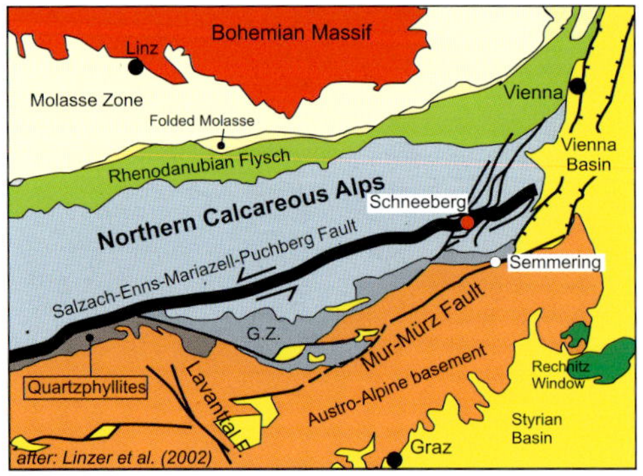

Schneeberg at the End of a Huge Fault Zone: In the last 30 million years, the Alps were successively pushed towards the east: the range was moved towards the Pannonian Basin along huge fault zones with strike slip displacement. The *Salzachtal-Ennstal-Mariazell-Puchberg Fault Zone* (in short: "SEMP") is one of the longest of these fault zones. It is about 400 km long and forms the northern boundary of the eastwards extruded part of the Alps. The rocks north and south of the fault zone were moved up to 70 km with respect to each other in a left lateral sense. Along the Salzach- and Enns Val-

leys the fault zone separates completely different rocks and tectonic units from each other. East of the township of Liezen in the Enns Valley the fault zone runs inside the Calcareous Alps. In the Gesäuse mountains it still parallels the Enns Valley, but further east the fault zone splits up into many branches and the displacement decreases. Just east of the Schneeberg (2076 m) the fault zones appears to merge into the delineating structures of the Vienna Basin, but geologists find it difficult to trace its further fate there. In the Schneeberg region the shattered rocks give clear evidence of its activity.

Left: Fault zones like this one near Talhof in the Semmering region are widely distributed in the Schneeberg region. They all form part of the huge system of the "Salzachtal-Ennstal-Mariazell-Puchberg Fault Zone" (photo: W. Kurz).

Right and below: A view of the Fohnsdorf Basin from the east. The basin is one of the largest Miocene pull apart basins of the Eastern Alps. In the background the Seetaler Alps. The diagram shows a north-south cross section through the basin.

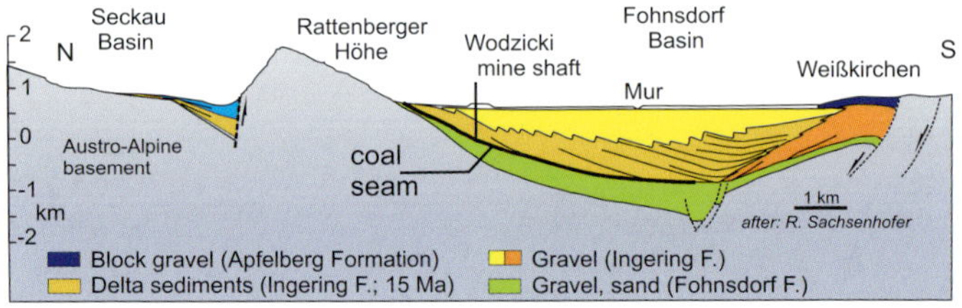

The Mur-Mürz Fault Zone: Together with the Lavanttal Fault Zone, the Mur-Mürz Fault Zone forms the most seismically active region of the Eastern Alps (p. 119). Both fault zones are oriented at roughly right angles to each other (p. 108) and run along major valleys. The fault zones formed in the Miocene in response to the interaction of the eastern end of the Alps with the Adriatic plate to the south and the *Pannonian Basin* to the east. Around 17 Ma the northward push of the Adriatic plate (at a rate of about 4 mm per year) separated the region south of the Mur-Mürz Fault Zone and east of the Lavanttal Fault Zone from the remainder of the Alps. This so called *Styrian Block* was now squeezed eastward and tilted slightly in the process. The Styrian Block includes the *Koralpe, Gleinalpe, Fischbach Alps* and the *Grazer Bergland.* The eastward movement of the Styrian Block opened a large pull-apart basin at its western corner: the Fohnsdorf Basin.

During the left lateral motion along the Mur-Mürz Fault Zone a series of other small pull apart basins also formed directly along this fault zone (p. 108). The *Leobener Basin*, the *Krieglach Basin*, the *Parschlug Basin* and the *Vienna Basin* in the northeast (p. 134). In all of these basins Miocene coal deposits formed. Volcanic tuff layers in between the sediments have been used to date the motion along the fault zones exactly.

The valleys of Mur, Mürz and Lavant are also of interest to geologists because they are the only major Alpine valleys that never contained glaciers in the ice ages and were therefore not re-carved in the last two million years. Therefore, the shape of the valleys, remnants of old river terraces and the palaeosurfaces around the valley can be used to interpret the Miocene landscape of the Eastern Alps.

Right: Flying along the upper Mürz Valley towards Vienna, the Mur-Mürz Fault Zone can be seen. Most of the earthquakes of the Eastern Alps occur here. In the middle ground the city of Mürzzuschlag. Behind Mürzzuschlag the fault zones runs towards the Semmering Pass which may be seen in the right hand background. On the right side of the pass is the Sonnwendstein (1523 m), and behind it the Vienna Basin. Left of the photo centre are Rax (2007 m) and Schneeberg (2076 m). Both are part of the Calcareous Alps. However, the sides of the Mürz Valley are made of coarse-grained gneisses of the Austro-Alpine basement. To the right of the freeway, one of the major Austrian natural gas pipelines may be seen.

Wörthersee is the largest (20 km²) and best known of the Carinthian lakes in southern Austria. Glacial gravels around its shores (yellow on the sketch) indicate that its valley was carved during the ice ages. The hills in the background are the Sattnitz Plateau (orange on the sketch). In the far background the Karavanke Range.

Carinthian Lakes: Carinthia is the southernmost Austrian province and offers something of every part of the geology of the Eastern Alps: there are the Penninic rocks of the Tauern Window, the Paleozoic rocks in the Karavanke Range and glacial geomorphology around the Carinthian lakes.

During the ice ages the Drava Glacier covered the entire region between Gurktal and the base of the Karavanke Range. In the east it reached as far as Griffen and Bleiburg where its terminal moraine may still be seen. The glacier carved deep into the garnet mica schists of the Upper Austro-Alpine basement (brown on map at right) that lies under the valley. After the retreat of the glaciers some 8000–10000 years ago, the valleys of the large Carinthian lakes were left behind. It is for this reasons that there are no lakes east of Griffen. Many of the smaller round lakes (e.g. Magdalen Lake near Villach) formed around dead ice bodies: blocks of ice that remained in the wake of the retreating glacier and then melted slowly and

statically. The chain of hills between Wörthersee and the valley of the Drau – the Sattnitz Plateau – is made up of Neogene conglomerates: solidified sediments deposited by rivers that came out of the Karavanke Range long before the ice ages in the Neogene (orange on the sketch at right).

Right: Geological sketch of the photo above. Brown = Austro-Alpine basement; light yellow = glacial moraine deposits; orange = Sattnitz Conglomerate; blue = Karavanke; light blue = Julian Alps.

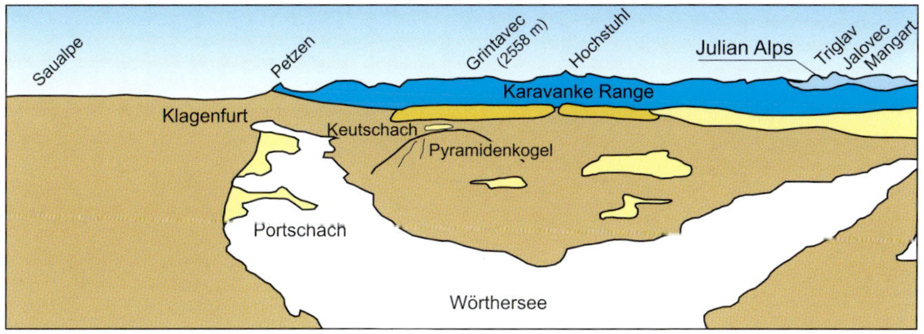

The Reißeck Massif (2965 m) at the southeast corner of the Tauern Window (red in the sketch at left) near the Millstätter lake in the Austrian province of Carinthia looking west. It may be seen how the Austro-Alpine rocks (brown on sketch) lie on both the south side (at left) and east side (photo centre) of the Tauern Window and strike down into the Möll Valley. The dome of the Tauern Window lies in its centre. The shear zone along which the Austro-Alpine rocks were dragged from the window is the Katschberg detachment. The glaciated peak in the centre of the image is the Hochalmspitze (3360 m). It lies in the central Tauern Window. Left behind it is Großglockner. In front of it in the foreground a block glacier. The Malta Valley lies in the right background. Behind it the Große Sonnblick (3030 m), Schober and Reitereck, all of which are part of the Tauern Window and are made up of rocks from the Penninic domain of the Alps. Only 30 Ma ago, these rocks were still buried deep in the crust below the Austro-Alpine nappes and heated to many hundreds of degrees. Only the motion along the Katschberg detachment brought the rocks to the surface in the last few million years. The Katschberg Pass itself may be seen on the very right of the photo. Behind it the Lower Tauern range of the Lungau region with its highest peak the Hochgolling (2862 m).

Above: The meteorological observatory on the Hohe Sonnblick (3 106 m) was built in 1886 onto the granitic "Zentralgneis" of the Sonnblick. It is the highest observatory in Austria.

Page 159: A view of the central Tauern Window from high above the Ankogel (3 252 m) looking west. The valley at right leads down to Böckstein – the northern end of the Tauern tunnel. The ridge at left includes the peaks Feldseekopf (2 864 m) and Weinflaschenkopf (3 008 m) of the Goldberg Group. The Möll Valley glacier can just be seen along this ridge in front of Großglockner. The famous resort of Sportgastein lies just out of sight in the bottom centre of the photo. Großglockner (3 798 m) – the highest peak of Austria – is the peak on the skyline in the photo centre. The sharp summit directly in front of it is the Hohe Sonnblick (3 106 m) with its observatory. The heavily glaciated dome to the right is Hocharn (3 254 m). The glaciated line of peaks leading to the right from Großglockner includes the Wiesbachhorn (3 564 m) near the right hand photo margin.

Below: Geological map of the Tauern Window.

The Tauern Window is the largest tectonic window in the Eastern Alps. Here, rocks from the Mesozoic European crust and the Penninic domain of the Alps penetrate the rocks of the Austro-Alpine. The window lies between Brenner Pass in the west and Katschberg Pass in the east. To the north it is bound by the Salzach Valley, and its southern boundary lies along a line from Spittal in Carinthia, across the mountains to Matrei in East Tyrol and then to Sterzing in the Italian province South Tyrol. The Tauern Window was opened about 20 Ma ago by gigantic detachment zones along the Brenner and the Katschberg. These shear zones dragged the cover nappes like curtains off the window to the west and east and exposed the rocks inside it.

Inside the Tauern Window, two different building blocks of the Alps are exposed. Similar to a cut onion, these building blocks include successively deeper crustal levels towards the centre of the window. These layers have the name "Schieferhüllen" (schist shells). The outermost shells are made from *Bündnerschiefer*: metamorphosed sediments from the former Penninic Ocean floor that occur widely in Switzerland. In the Tauern Window, these rocks belong tectonically to the Glockner Nappe. The Glockner Nappe also includes rocks from the Penninic oceanic crust itself. When this ocean was subducted, its rocks were pushed underneath the Austro-Alpine and are therefore now present as metamorphic schists. Near the East Tyrolean township of Matrei, some of these rocks are transformed into eclogites.

Below the Penninic nappes, coarse-grained orthogneisses are exposed in the central Tauern Window. Although they are the structurally deepest unit of the window, they built up the topographically highest peaks of the Hohe Tauern: *Ankogel*, *Hochalmspitze*, *Goldberg-*, *Sonnblick-*, *Granatspitz-*, *Venediger-* and *Reichenspitz Group*. All these mountains are made from deformed granites (orthogneiss). They are part of the Mesozoic European middle crust and are therefore equivalent to the External Massifs of the Central Alps. In Variscan times (380 – 310 Ma) these rocks intruded the crust as granites and today bear the name Zentralgneis (Central gneiss). In places, these rocks are covered by remnants of sediments that once formed the sedimentary cover of the European crust (Hochstegen limestone and Seidlwinkel-Trias).

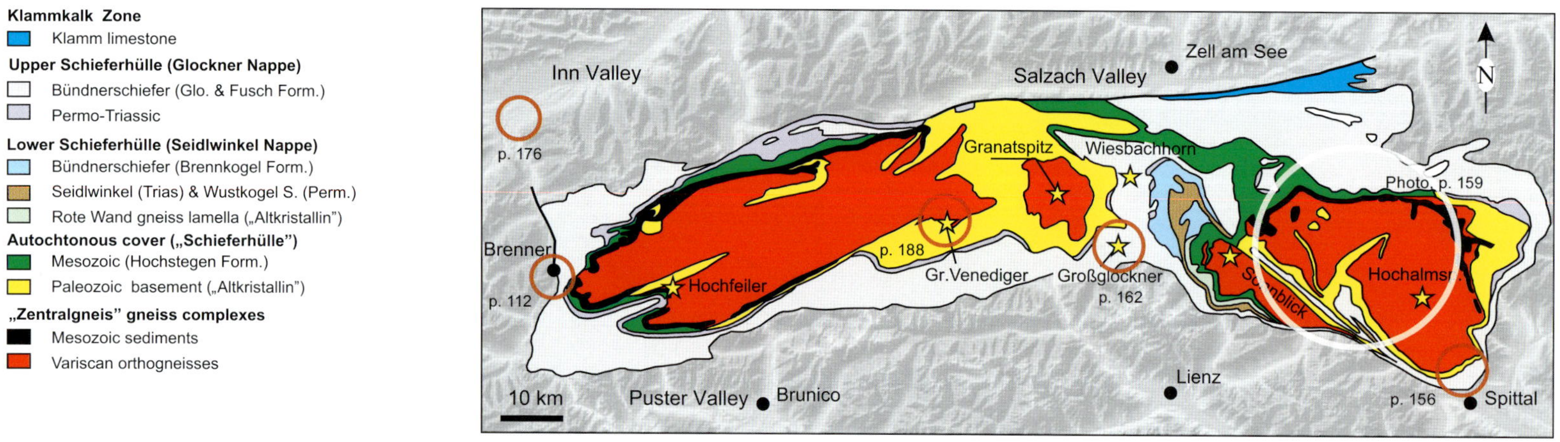

Page 161: The prominent Wiesbachhorn (3 564 m) seen from the north from high above the Fuscher Valley. The black rocky peak to the left of Wiesbachhorn is the Hohe Dock (3 348 m). Behind it to its right the Große Bärenkopf (3 396 m). To the left of Hohe Dock there are Großglockner and Fuscherkarkopf (3 331 m) in the left foreground. The triangular peak is the Sinwelleck (3 261 m). In the bottom left corner of the photo, the first switchbacks for the Großglockner Hochalpenstraße may be seen. About 500 vertical metres above the valley floor there is an obvious shoulder which is a remnant of the valley floor before the ice ages.

Below: The Moserboden dams in the upper Kaprun Valley are part of one of the largest hydroelectric schemes in the Eastern Alps.

Wiesbachhorn – Highest Relief in the Eastern Alps: Wiesbachhorn (3 564 m) is located along the northern margin of the Tauern Window south of Zell am See and is considered as the big rival to the Großglockner. The eastern flank of the mountain drops for about 2 500 vertical metres down into Fuscher Valley, thus forming the highest summit – valley profile in the Eastern Alps. Geologically, Wiesbachhorn and Großglockner both lie entirely within the Penninic rocks of the Glockner Nappe. Thus, ironically, the highest peaks of Austria are made from rocks that are derived from the bottom of the ocean. When this ocean floor was pushed underneath the Austro-Alpine nappes, the sediments from the ocean floor were transformed into garnet bearing calcareous mica schists and the basalts of the oceanic crust itself to prasinites. Prasinites are very erosion resistant rocks and they are partly responsible for the high elevation of Großglockner: a band of prasinites builds up the summit of the peak. On the northern slope of Wiesbachhorn it may clearly be seen how prasinites and mica schists lie like roof tiles on the more centrally located rocks of the Tauern Window. During uplift and exhumation of the Tauern Window in the last 20 – 30 Ma the layers were tilted steeply to the north and many kilometre-thick layers eroded. During the glaciation periods, the valleys on either side of the Wiesbachhorn were carved down for many hundreds of metres: the *Fuscher Valley* in the east and the *Kaprun Valley* in the west.

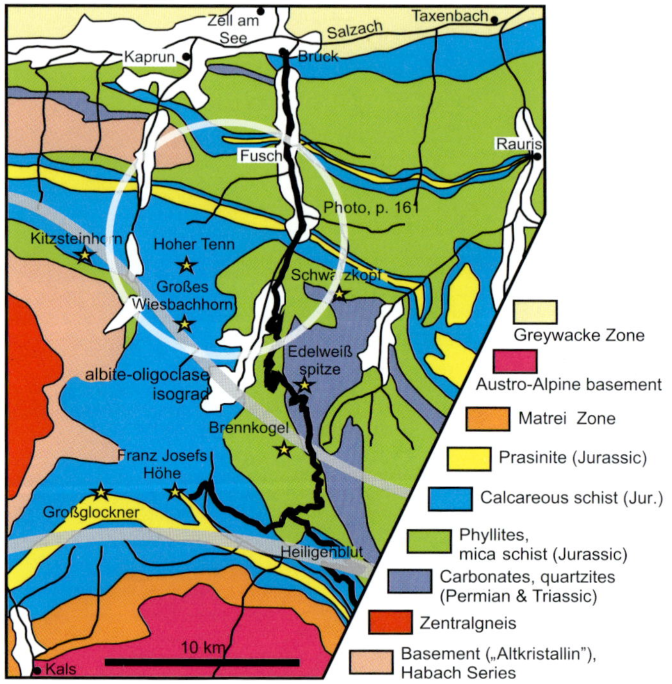

*Left and page 163: **Großglockner** (3 798 m) from the east. The largest glacier of the Eastern Alps – the Pasterze – lies at its base. The glacier runs down the Johannisberg (3 453 m) and measures 9 km to its terminus at an elevation of about 2 100m. The Gamsgrubenweg – a tourist walk along the Pasterze glacier – may be seen above the ridge in the foreground. The normal ascent route up Großglockner leads from the Pasterze leftwards up to the lower end of the Hofmannskees glacier and then in a large arc across the glacier to the Erzherzog-Johann Hut, which can be seen against the sky on the left hand shoulder of Großglockner. The peaks to the right of Großglockner are Glocknerhorn (3 680 m), Glocknerwand (3 721 m), Teufelskamp (3 510 m), Romariswandkopf (3 508 m) and – the snow pyramid at the very right – Johannisberg (3 453 m). At the very left on the skyline the Hochgall (3 436 m) – highest peak of the Rieserferner Group. Just below the Glocknerwand the famous ice climbing routes of the Großglockner region: Pallavicini couloir, Bergler couloir and Mayerl ramp. To its right is the horseshoe fall of the Pasterze glacier that separates the accumulation region from the ablation region of the glacier.*

Top right: Großglockner (at right) and Glocknerwand (at left) from the south. The prominent ridge that leads directly to the summit of Großglockner is the Stüdlgrat. The rapid melting of the glaciers has exposed a spectacular banding of greenish prasinites and brown-grey Bündnerschiefer of the Penninic Glockner nappe in the Glocknerwand.

Below: Geological north-south cross section through the Tauern Window in the Großglockner region, roughly parallel to the Großglockner Hochalpenstraße.

Großglockner (3798 m) is the highest peak of Austria and geologically located in the centre of the *Tauern Window.* As for its rival, the Wiesbachhorn, the mountain lies inside the *Schieferhüllen* ("schist hulls") of the Tauern Window and its rocks were therefore part of the Penninic Ocean floor in the Mesozoic. The mountain is made up of prasinites (former basalts from the ocean floor) and mica schists (former deep sea sediments) of the *Penninic Glockner Nappe* (p. 160). The rocks were metamorphosed about 30 Ma during the Alpine metmorphic events.

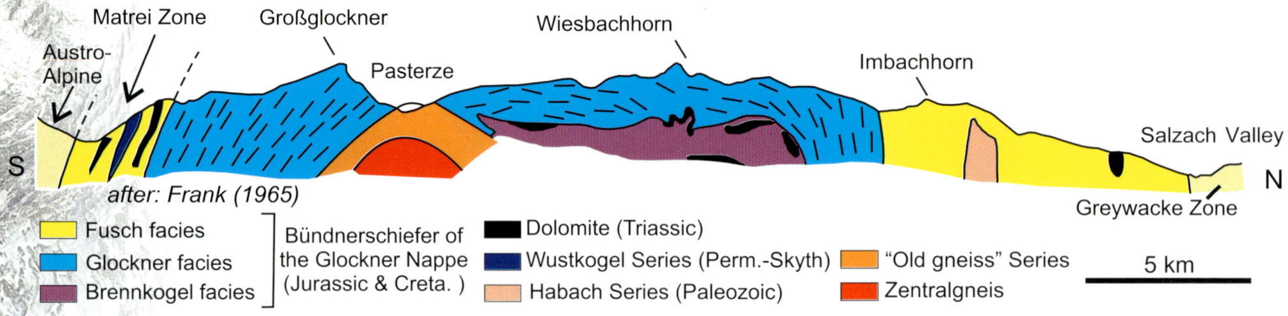

Austrias largest icefield in the Großglockner region: the Pasterze. The glacier covers roughly 17 km² in area and 2 cubic kilometres of ice. In 1850 it still covered almost twice this area. Other glaciers of the Großglockner region are the *Hofmannskees*, the *Kleinglocknerkees* and the *Glocknerkees*. The image on this page shows the Pasterze region from high above the Johannisberg looking towards Wiesbachhorn and Hohe Riffl in the middle ground (3 338 m). The lakes in the left hand part of the image are the Moserboden dams that belong to the Kaprun hydroelectric scheme. The prominent Wiesbachhorn is in the centre of the photo. Großglockner at the right hand photo margin. The rocks in the left hand face of the Hohe Riffl show the banding of calcareous mica schists and greenish prasinites of the Penninic Glockner nappe.

Left: Comparison of the ice levels of the Pasterze glacier in 1850 and 2002 (after: Kellerer-Pirklbauer, 2008).

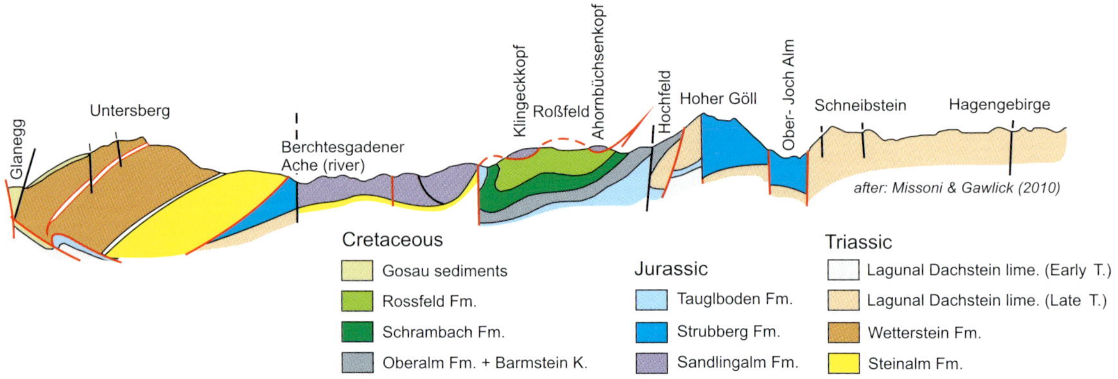

Watzmann (2713 m) dominates the valley of the Königssee Lake in the Berchtesgadener Alps. In the photo, the township of Berchtesgaden may be seen directly behind the lake at the base of the flat-topped Untersberg. Behind Untersberg lies the city of Salzburg.

Geologically, Watzmann, the Steinerne Meer Massif, the Hagengebirge and the Untersberg are all made up of Triassic

Dachstein limestone (210 Ma). At its base this limestone lies on the somewhat older Hauptdolomit (Ramsau dolomite). The typical banking of the Dachstein limestone is clearly visible in the Watzmann east face, where it was tectonically tilted by about 30° to the northwest. A typical indicator fossil for the Dachstein limestone is the *Kuhtrittmuschel*. The reddish colouring of the limestone indicates traces of iron. As with any other limestone, Dachstein limestone is water soluble and therefore karstifies easily. As a consequence, countless caves have formed in the massifs in the youngest part of their geological evolution.

Königssee is also the location of one of the largest historic landslides of the Alps, which cut the lake into two separate parts in the 12th century: the *Obersee* (on the photo in the foreground in the shade) and the *Königssee* itself.

Above: *The north faces of the Wilde Kaiser Massif before a summer storm.*

Below: *Kitzbühel, the legendary Tyrolean city of sports, looking east. Behind the cable car at the right hand margin of the photo is the Ochsenhorn (2 511 m, highest peak of the Loferer Steinberge). The limestone peaks on the skyline are the Leoganger Steinberge with their highest peak the Birnhorn (2 634 m).*

The massif of the Wilde Kaiser and its neighbour, the *Zahme Kaiser,* lie between Kufstein and St. Johann in Tyrol just east of the Inn Valley. The massif contains 20 peaks above 2 000 elevation. The highest peak is the *Ellmauer Halt* (2 344 m). Akin to many other peaks of the Calcareous Alps, the entire rocky part of the mountain above about 1 400 m is made up of a massive slab of Wetterstein limestone and dolomite, which was deposited around 220 – 240 Ma into the Tethys Ocean. A thin band of Raibl beds separates these rocks from the lower parts of the mountain. This lower part is made up of Hauptdolomit (deposited 210 Ma). Tectonic movements during the Jurassic and Cretaceous tilted the sedimentary layers in many places so that the younger Hauptdolomit lies below the somewhat older Wetterstein limestone. This tilting and much later exposure by erosion are also responsible for the formation of the vertical rock needles and steep slabs that are well known to climbers.

Near Scheffau and Ellmau, south of the Wilde Kaiser, rocks of the *Werfen Formation* are exposed, and south of there the Greywacke Zone of the Alps builds up much of the *Kitzbüheler Alps. Paleozoic dolomites,* volcanic porphyries and mafic schists are the dominant rock types there.

The flanks of the Wilde Kaiser Massif are also well known for their glacial history. A complicated suite of moraines and glacial terraces formed when sidearms of the Inn Valley glacier were pushed up the valley to the east.

Page 169: Ackerlspitze (2 329 m) is one of the highest peaks of the Wilde Kaiser Massif. Other peaks on the photo are the Regalspitze (2 253 m, to the right along the ridge); Mittlerer Kaiser (2 011 m) and Torspitze (2 204 m, foreground at right). On the skyline at right the Große Wiesbachhorn (3 564 m), and in front of it the Kitzbüheler Horn (1 996 m). To the right behind Ackerlspitze the Hochkönig Massif (p. 172).

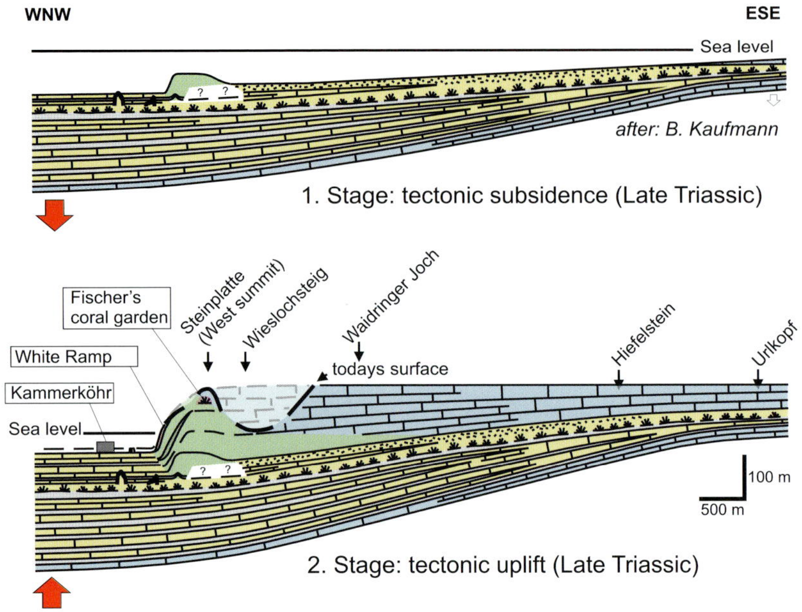

The Steinplatte (1 896 m) near the Austrian town of Lofer is part of the *Chiemgauer Alps* located in the border region between Germany and Austria. The slab-shaped mountain offers a unique display of a Triassic (200 – 250 Ma) reef that is unparalleled worldwide. As part of the Northern Calcareous Alps this reef grew somewhat south of the equator along the northern margin of the Meliata-Hallstatt Ocean (Meliata Hallstatt is the name for the westernmost embayment of the Tethys Ocean).

The Steinplatte reef is the continent directed end of a carbonate platform that reached through the *Loferer-* and *Leoganger Steinberge* and the *Steinerne Meer* to *Hochkönig* at the oceanward end of the platform.

Minor tectonic lowering and uplifting of the surface during the reef development in the Upper Triassic created a shallow basin between the carbonate platform and the continent, similar to the basin formed today between the Australian barrier reef and the Australian coast, or between the Bahamas and the Florida coast. This basin is called the Kössen Basin. The Steinplatte was located along the eastern margin of this basin. In the vicinity of the Kammerköhr restaurant, there is the steep (30°) slope from the tropical shallow sea (near the Steinplatte summit and the entire region east of there) into the Kössen Basin close to land. The coral limestones in the summit region of the Steinplatte are amongst the first "modern" coral reefs in the world. They are made of a coral family called *scleractinia*, which builds up most of the coral reefs around the world today. Because of the unique outcrop conditions, the Steinplatte has been a playground for the international research projects of geologists and palaeontologists for the last 15 years.

Above and at right: Schematic illustration of the reef development at Steinplatte. The colours correpond to the diagram on p. 172.

Page 171: Steinplatte seen from the west. In the Triassic, the obvious face in the foreground formed the reef slope towards the Kössen Basin. Behind it begins the carbonate platform that reaches all the way to Hochkönig. The Steinerne Meer massif may be seen on the skyline.

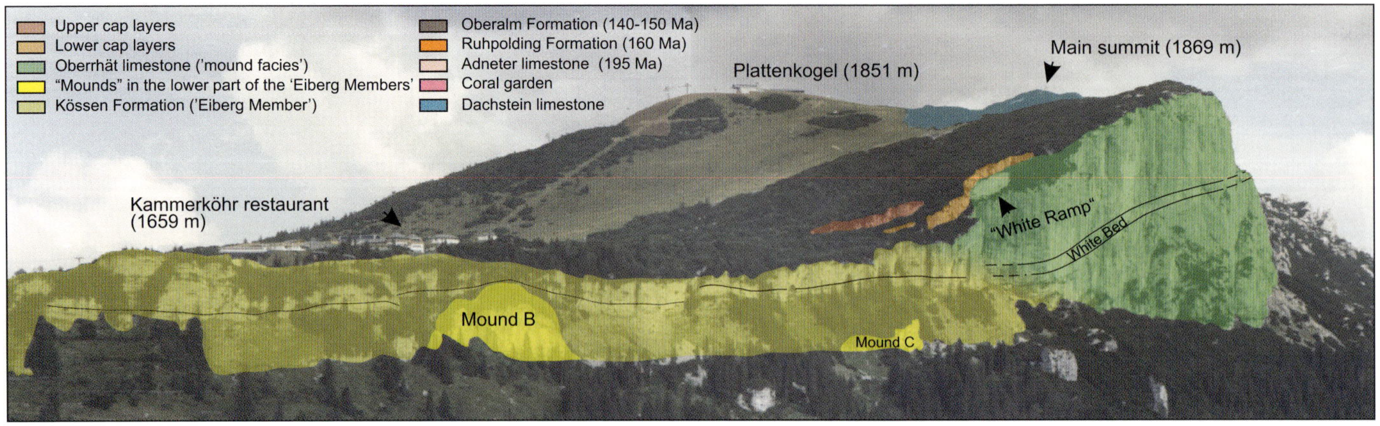

Above: The Übergossene Alm in the summit region of the Hochkönig Massif.

Right: The Torsäule peak seen from the summit region of Hochkönig looking east.

Below: Schematic sketch of the carbonate platform of which Hochkönig forms the eastern end (ring = photo, p. 173).

Page 173: Hochkönig just before a summer storm seen from the east. The hut on the summit is the Matrashaus. In the bottom left corner of the photo the Arthurhaus may be seen.

The Hochkönig (2941 m) is the southeastern-most mountain of the Steinerne Meer Massif near the Austrian city of Bischofshofen. As part of the Calcareous Alps it forms the eastern end of an enormous carbonate platform that lay in the Triassic along the northern margin of the Tethys Ocean. East of the mountain, the Hallstatt Basin follows, in which the Tethys Ocean was somewhat deeper. To the west, the carbonate platform reached as far as the Steinplatte (p. 170). The summit region of Hochkönig is made up of massive, karstified Dachstein limestone (200 – 210 Ma).

The thick slab of Dachstein limestone lies on top of the more brittle Hauptdolomit (below the level of the Torsäule). Below the Hauptdolomit, Raibl beds, Wetterstein dolomite (225 Ma) and Gutensteiner limestone follow. On the level of the Arthurhaus lodge (about 1 500 m), the entire carbonate succession lies on top of the soft layers of the schists of the Werfen Formation (250 Ma). The landforms abruptly change from rugged limestone cliffs to rounded hills.

The sequence of rocks resembles that of many other regions of the Calcareous Alps, for example the Gesäuse mountains (p. 140) or the Hochschwab region (p. 142). At the base of Hochkönig Massif, the schists of the Werfen Formation lie on top of Paleozoic rocks of the Greywacke Zone that builds up most of the region south of the city of Bischofshofen.

The summit plateau of the Hochkönig is covered by a small rapidly melting glacier that gives the plateau its name: *Übergossene Alm*. In the rocks of the Werfen Formation near the Arthurhaus lodge there is the oldest copper mine of the Alps. Copper has been mined here since about 3700 years ago. The mine was finally closed down in 1977.

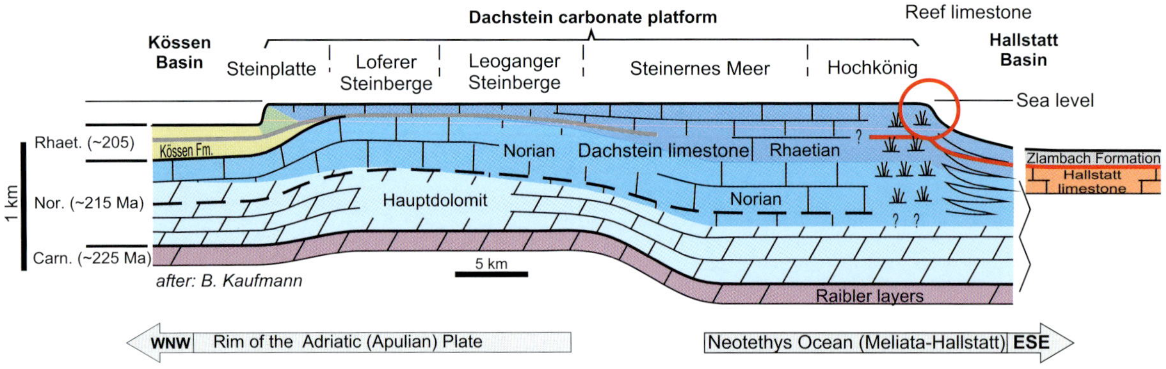

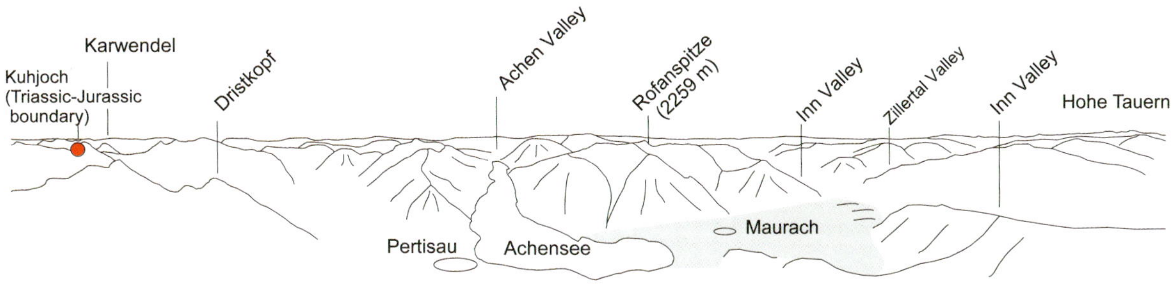

Achensee from the southwest. The panorama covers 270°. The red dot on the sketch shows the location of the internationally accepted type locality for the Triassic-Jurassic boundary. It lies on the Kuhjoch on the northern slope of Fleischbank (2026 m) in the Rissbach Valley of the Karwendel mountains. The grey shaded region shows the scree fan that blocks the drainage of the Achensee towards the Inn Valley.

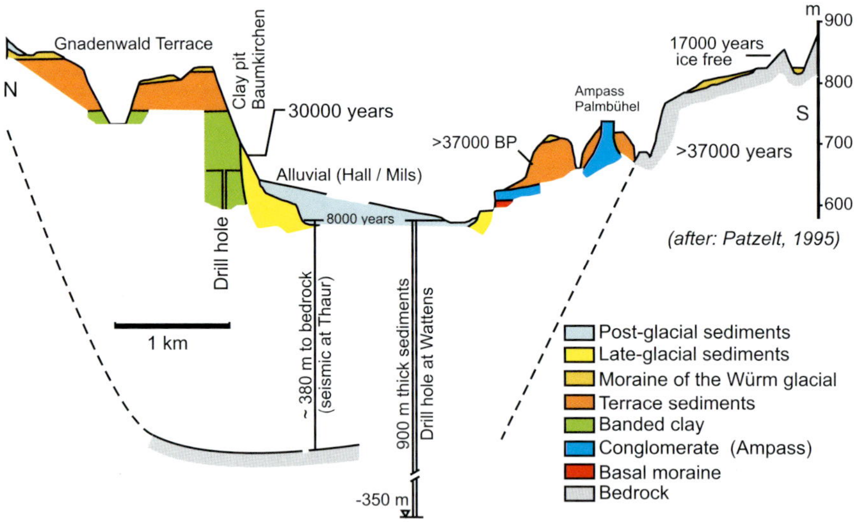

Above: Geological cross section through the glacial terrace landscape near Innsbruck.

Right: The Patscherkofel (2 246 m) is the local mountain of Innsbruck. Up to an elevation of about 1 800 m its northern side is made up of quartzphyllites of the Lower Austro-Alpine. The summit region is made up of mica schists and gneisses of the Upper Austro-Alpine.

Page 177: The city of Innsbruck seen from the south from high above the Patscherkofel. The Innsbrucker Mittelgebirge with the Lanser Köpfl is clearly visible in the foreground. The Martinswand is the vertical cliff directly above the valley floor west of Innsbruck. The morphology of the entire Inn Valley was formed by glaciers. Remnants of moraines are preserved in many places around the valley.

The valley of the Inn is largely located along a geological strucutre called the Inntal Fault. This is one of the large strike slip zones along which the central part of the Alps was pushed eastwards during the Miocene (p. 108). The motion along the fault juxtaposed very different packages of rock: on the north side of the Inn Valley there are the Northern Calcareous Alps including the *Wetterstein Gebirge*, *Karwendel* and the *Rofan*. South of the Inn Valley there are the *Ötztaler*, *Stubaier*, *Tuxer* and *Kitzbüheler Alps*, all of which are made up of metamorphic rocks of the Austro-Alpine basement.

The Inn Valley is also geologically known for its complicated glacial evolution during the youngest part of its geological history. During the ice ages the Inn glacier used the zone softened by deformation along the Inn Valley fault zone to carve the valley several hundreds of metres deeper. As a consequence, many tributaries were redirected, for example those in the region of the Achensee (p. 174). Moraines from this time are preserved in many places. The deep U-shaped valley carved by the glaciers reached many hundreds of metres deeper than todays valley floor. However, the valley was already refilled by gravels and other sediments during the ice age. Vertical cliffs like the Martinswand are only the upper half of this gigantic U-shaped valley. The *Höttinger Breccia* is a solidifed screeslope from the Riß-Mindel-interglacial from about 200 000 years ago that is a popular building stone in the Tyrolean capital of Innsbruck.

The Karwendel mountains in the region of the Rappenspitze west of Achensee. In the right hand middle ground the Sonnjoch (2457 m). The chain of mountains at left includes the two highest peaks of the Karwendel: the Birkkarspitze (2749 m) and Ödkarspitze (2738 m). Along the base of the mighty north faces, there is a 100 m thick band of banked limestone: Alpine Muschelkalk (240–245 Ma). Above this the Wetterstein limestone (230–240 Ma). The thrust zone between the Inntal Nappe and the Lechtal Nappe is located on the meadows just below the Muschelkalk band.

The Karwendel Mountains lie on the north side of the Inn Valley between the Rofan- and Wetterstein Massif. The best known part of the Karwendel Mountains is the *Innsbrucker Nordkette* – the chain of mountains that rises steeply above the Tyrolean capital for almost 2000 vertical metres. Geologically, the Karwendel is part of the Northern Calcareous Alps that were deposited in the Triassic along the northern margin of the Tethys Ocean on the shelf of the Adriatic (Apulian) plate. During stacking of the crustal fragments in the Jurassic and Cretaceous, the different parts of the Tethyan sediments were pushed from southeast to northwest, forming what is known as the Inntal Nappe and the Lechtal Nappe. The topographic arrangement of the Karwendel Mountains into several east-west striking ranges is the result of this transport. Between the *Innsbrucker Nordkette,* the chain of the Bettelwurf (2796 m) and the chain around the Birkkarspitze (2749 m), the Wetterstein limestone is folded into huge open east-west striking folds. At the bottom of these folds Raibl beds are exposed. North of Birkkarspitze, the Inntal Nappe lies tectonically on top of the Lechtal Nappe, which, in turn, lies on the Allgäu Nappe.

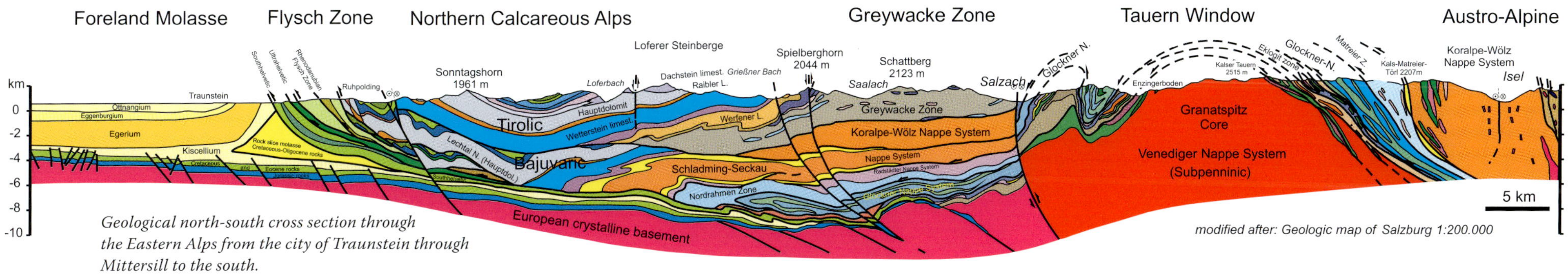

Geological north-south cross section through the Eastern Alps from the city of Traunstein through Mittersill to the south.

The Zugspitze (2 962 m) is the highest mountain of Germany and the highest peak of the Wetterstein Massif. The enormous 800 m high summit build up is mainly made up of Wetterstein limestone. This rock type occurs widely along the entire chain of the Northern Calcareous Alps. The rock is predominantly made up of algae and was deposited in the middle Triassic (230 Ma) in lagoons that formed on the slowly submerging Adriatic plate. Wetterstein limestone is somewhat older than the upper Triassic carbonates that occur in the Alps: Hauptdolomit and Dachstein limestone. The photo shows the west faces of the Zugspitze Massif looking east. Directly behind the summit of Zugspitze lies Innsbruck at the intersection between Brenner- and Inn Valleys. The crystalline basement of the Hohe Tauern, Tuxer Alps and Stubaier Alps form the peaks on the skyline.

Allgäu Alps: West of Innsbruck, north of the river Inn and east of the Rhine Valley there are a series of mountain ranges that cover the transition between two Mesozoic continents: the *Lechtal Alps*, *Allgäu Alps*, *Rätikon* and *Bregenzer Wald*. The southernmost of these ranges are the *Lechtal Alps*. They are entirely part of the Austro-Alpine Calcareous Alps and include the mountains between the rivers Lech and Inn (p. 184).

The *Allgäu Alps* north of the river Lech are similar to the Lechtal Alps in their southern part. There they are made up of Hauptdolomit. As such, they are tectonically part of the Austro-Alpine that once was part of the Adriatic plate south of the Penninic Oceans. North of the Walsertal, the Allgäu Alps are made up of marly deep sea sediments that once were part of the Valais Ocean in the Mesozoic and now are part of the Penninic Flysch Zone. The rocks are soft and tend to form landslides. Between Dornbirn, Sonthofen and Forggensee the Flysch Zone forms steep summits with grassy flanks (e.g. Üntschenspitze, 2135 m). A broad band of the so called Faltenmolasse forms the hilly landscape north of the Flysch Zone.

All the way in the west is the *Bregenzer Wald*. Here limestones re-appear north of the Flysch Zone. West of the Rhine Valley, these limestones merge with the Helvetic nappes of Switzerland. Between Feldkirch and Oberstdorf the Helvetic limestones build up the summits of Kanisfluh (2044 m) and Winterstaude (1844 m). These mountains form *klippen* on top of the Flysch rocks. Tectonically, these limestones are very different from the limestones in the Allgäu Alps: in the Mesozoic they were deposited onto the European plate north of the Penninic Ocean.

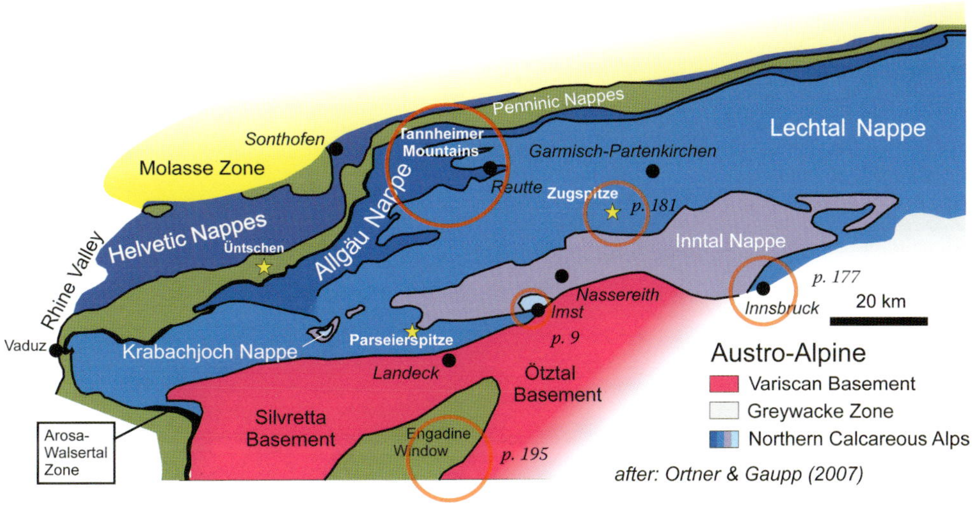

180° panorama of the Lech Valley from high above the township of Weißenbach am Lech. The carbonate massifs in the centre of the image are the Tannheimer Mountains with the Gimpel (2 176 m), Kellenspitze (2 238 m) and Gehrenspitze (2 163 m). To the right behind them Forggensee and the city of Füssen, in the foreground Höfen (left of the river Lech) and Reutte (further in the back on the right side of the river). Left of the image centre is the Haldensee in the Tannheimer Valley. Between Tannheimer Valley and Lech Valley (at left) the view direction is towards the west onto Hochvogel (2 592 m), Lailachspitze (2 270 m) and Gaishorn (2 247 m). Behind them the valley of Oberstdorf. Lake Constance can be seen on the skyline (largest ring = photo above).

The Lechtal Alps are bound by the valley of the Lech to the north, the valley of the Inn and the Verwall Group to the south and the Lechquellengebirge to the southwest. To the east, the Lechtal Alps are neighboured by the Wettersteingebirge (p. 180). Both are part of the Northern Calcareous Alps and are geologically very similar, but the Lechtals Alps feature a larger variety of the Mesozoic sedimentary succession. In part, this is because the Lechtal region was tectonically broken up into three separate nappes that are partly folded into each other: the northern Allgäu Nappe, the central Lechtal Nappe and the southernmost (and uppermost) Inntal Nappe. In many of the mountains of the Lechtal Alps, Triassic Wettersteinkalk limestone forms the base and Hauptdolomit, Juarassic marls and Cretaceous limestones form the upper parts of the peaks. The highest peak of the Lechtal Alps, the Parseierspitze (3 036 m), is the only mountain of the Northern Calcareous Alps that exceeds the 3 000 m mark. It is made up of limestone that is folded and mantled by radiolarites in the summit region. The Leiterspitze (2 750 m), Dremelspitze (2 733 m) and many of the other peaks are built up of shattered Hauptdolomit. In the Holzgauer Wetterspitze (2 895 m) a massive limestone occurs that corresponds to the Dachstein limestone.

Above: A view of the Forggensee in the northern foreland of the Alps seen from high above the Aggenstein (1 986 m). Aggenstein is part of the Tannheimer mountains of the Allgäu Alps and belongs to the Northern Calcareous Alps. The chain of hills in the middle ground is also part of the Calcareous Alps. Behind it there is a narrow band of flysch rocks and behind those the rocks of the Molasse Zone.

Right: Geological north-south cross section from the town of Sonthofen (in Bavaria) in direction of the Arlberg Pass in Austria.

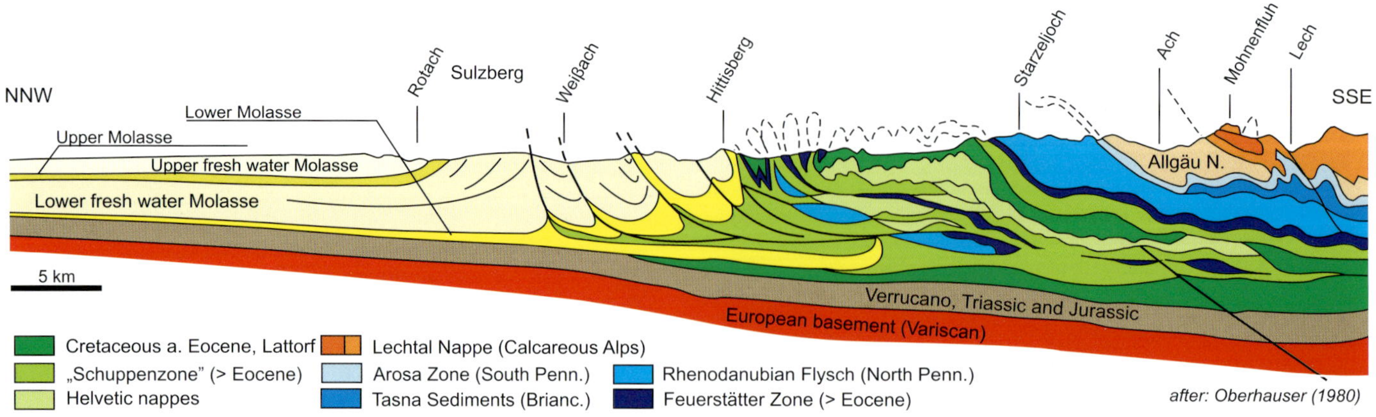

An overview of the Lechtal Alps looking southwest. In the foreground at right the Lech Valley. The glaciated peak in the right half of the photo is the Hohe Riffler (3 168 m), highest peak of the Verwall Group. The highest peak in the left part of the photo is the Parseierspitze (3 036 m), highest peak of the Northern Calcareous Alps. The double summit behind it is the Fluchthorn (3 399 m, Silvretta Group). On the skyline the Bernina Group.

Right: Three dimensional diagram illustrating the relationship between South Alpine (red), Austro-Alpine (brown) and Tauern Window (grey) (red rectangle = map below).

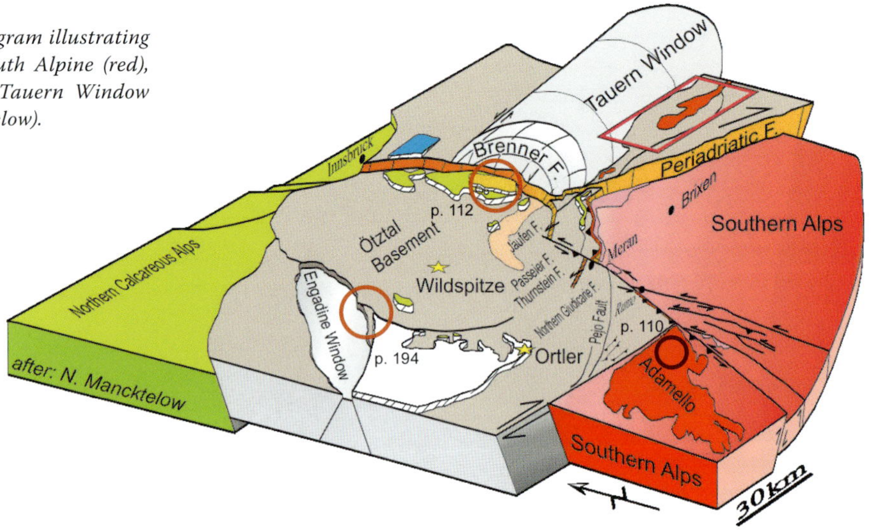

Right: Geological map of the Austro-Alpine between Periadriatic Line and Tauern Window. The northern black line is the southern boundary of the Tauern Window. The southern line is the Deffreggen-Antholz-Line (DAV). The Periadriatic Line itself lies just outside the southern map boundary (ring = photo, p. 187).

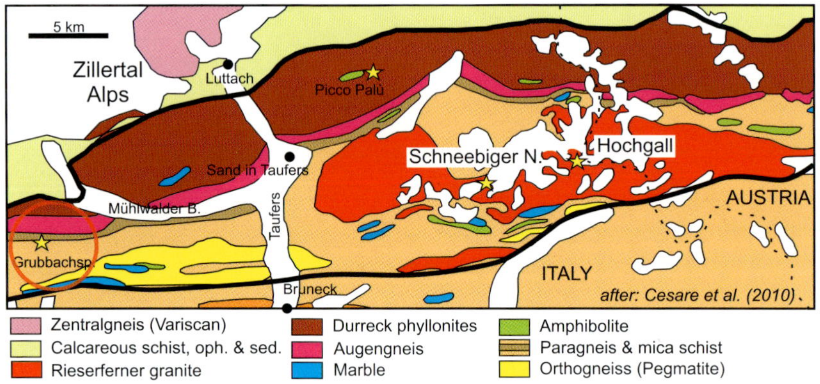

The Austro-Alpine between Tauern Window and Southern Alps: In the Puster Valley of South Tyrol, the three major tectonic parts of the Alps occur all next to each other: the Penninic rocks of the Tauern Window, the South Alpine and the Austro-Alpine. The Periadriatic Line that separates the Austro-Alpine from the South Alpine runs along the northern side of the Puster Valley from Sillian through Bruneck, crosses the Brenner Valley south of Sterzing and then arcs through Meran to the Adamello Massif in the southwest. The boundary of the Tauern Window leads from Matrei in East Tyrol through the township of Luttach in the Ahrn Valley to its southwest corner near Sterzing south of the Brenner Pass. Between the Tauern Window to the north and the South Alpine to the south there is a narrow wedge of gneisses of the Austro-Alpine basement. The *Villgratener Mountains,* the *Rieserferner Group* and the southern part of the *Zillertal Alps* are part of this wedge. Near Sillian, this wedge is about 30 km wide, but near the township of Mauls southeast of Sterzing, this wedge of Austro-Alpine rocks is only 3 km wide. Two granitic intrusions lie inside this wedge, which both are part of the chain of Periadriatic intrusives that intruded the Alps about 35 Ma (p. 110): the *Rensen Granite* and the *Rieserferner Granite.* The Rensen Granite is about 1.5 km wide and 6 km long and is located in the Valser Valley north of Mühlbach. The Rieserferner Granite ist 4.5 km wide and 40 km long and includes the highest peak of the Rieserferner Group: Hochgall (3436 m).

Page 187: The Pfunderer Mountains along the southern margin of the Zillertal Alps on the north side of the Puster Valley near Terenten. The peak in the foreground is the Grubbachspitze (Monte Gruppo, 2809 m). The lake at right is the Nevesstausee at the end of the Mühlwalder Valley. The highest peak on the skyline is Hochfeiler (3510 m), the highest mountain of the Zillertal Alps. To its right the Hohe Weißzint (3371 m). At the right photo margin the Große Mösele (3478 m).

Großvenediger (3657 m) is the highest peak of the Venediger Group in the *Hohe Tauern* range of Austria. It forms a majestic pyramid some 30 km west of Großglockner. Geologically, Großvenediger lies along the southern margin of the *Zillertal Kern*, one of the four *Zentralgneis massifs* of the Tauern Window (the other three are the *Granatspitz Kern*, *Sonnblick Kern* and *Hochalmspitz Kern*) (p. 158). These four granite and gneiss massifs formed during Variscan times more than 300 Ma ago and once were part of the European middle crust. They correspond to the *External Massifs* of the Central and Western Alps. Just south of Großvenediger, these gneisses are in contact with garnet bearing mica schists that are also of Variscan age, but are considered to be part of the Paleozoic cover ("autochthonous Schieferhülle").

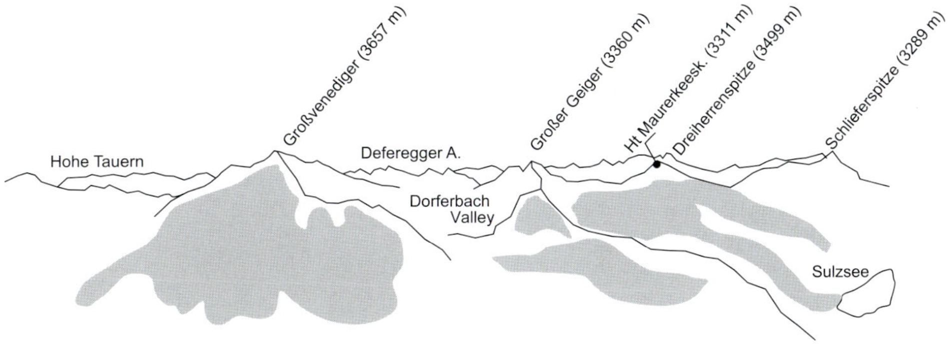

Großvenediger from the west. The mountain is made up of orthogneisses that formed more than 300 Ma ago. The Große Geiger to the right of Großvenediger is made up of garnet mica schists that surround the Venediger gneiss massif.

Page 191: The landslide of Köfles in the Ötz Valley in Tyrol. In the background the Tschirgant landslide at the Ötz Valley entrance. The colour scheme of the sketch corresponds to the geological cross section at right.

Below: Numerical model of the stresses in valley flanks during and after a glaciation period. It may be seen that highest surface stresses occur after the melting of the ice in the middle of the slope (black square).

The three maps below at right show the sliding mass just before the catastrophic event, just after and today. Black dot = centre of gravity of the sliding mass (after: Brückl et al. 2001).

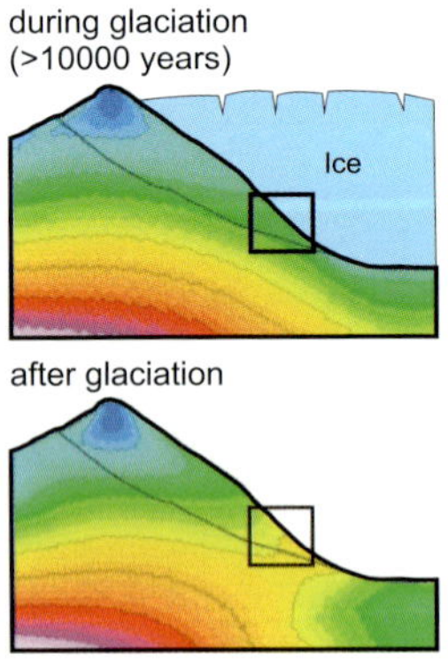

The Landslide of Köfels: About 10000 years ago, the last ice age left behind oversteepened U-shaped valleys in the Alps with flanks that were no longer supported by ice. This led to stresses in the rocks of the flanks and many catastrophic landslides followed. One of the largest was the landslide of Köfels in the Central Ötz Valley west of Innsbruck. In fact, the Köfels landslide is the largest of its kind in the Alps that occurred within the crystalline basement. About 8700 years ago more than 3 km^3 of rocks slid at up to 200 km/h from the western slope into the valley. On the opposite side of the valley, the momentum of the impact literally shattered the rocks. The upper half of the sliding mass then skidded on top of the lower half further into the Horlach Valley – a side valley of the larger Ötz Valley opposite the landslide. The river in the valley, the Ötztaler Ache rapidly formed a kilometre long lake that later filled the entire basin of Längenfeld with lake sediments until the river carved a new valley through the sliding mass: the *Maurach* Gorge. Because of the finely ground rocks, this happened very rapidly. The landslide is famous for its findings of pumice. For a long time volcanoes were sought in the region and even a meteorite impact was postulated for the region because of these pumice findings. Today it is known that

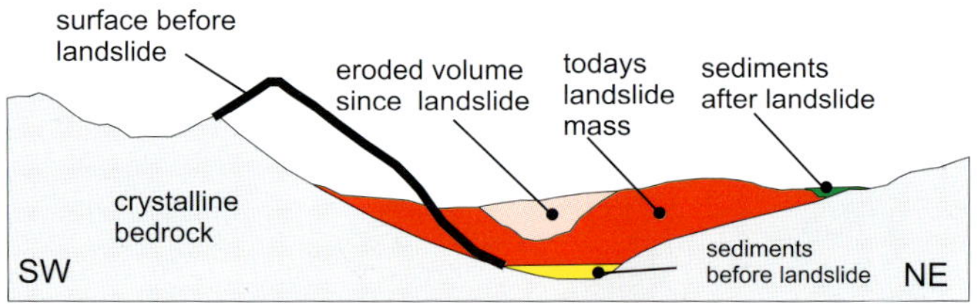

this rock (sometimes referred to as: *hyalomylonite*) formed as a foamy product from frictional heating of the sliding rocks mass.

The landslide is also well known for the fact that the nearby township of Umhausen has an above-average level of radon in its ground water. This high level of radon could be connected with the finely ground rock mass that constitutes the bulk of the sliding mass.

Aside from the Köfels landslide there is another of the large landslides of the Alps in the Ötztal region: the Tschirgant landslide occurred 3000 years ago and its mass forms a moraine-like region where the Ötz Valley meets the Inn Valley.

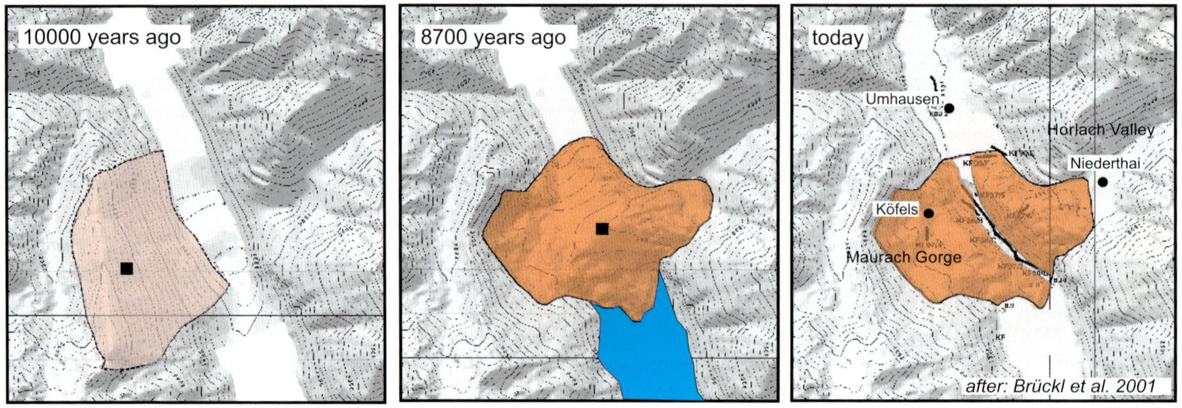

Above: The photo shows the southern margin of the Ötztal Alps looking south into the Passeier Valley from directly above the Timmelsjoch pass. Geologically, the view is onto the Schneebergerzug (orange on the sketch right), that lies here within the Ötztal basement (red). The mountains south of Meran lie beyond the Periadriatic Line. They are part of the Southern Alps. The mountains there consist of Bozen quartzporphyry.

Right: Sketch of the photo with a schematic view of the geology.

The Ötztal Alps lie west of the Stubai Alps between Inn Valley to the north and Adige Valley to the south. The Ötztal Alps are the largest region of the Eastern Alps that lies higher than 3000 m above sea level. In the mean, the Ötztal Alps are even higher than the Bernina or the Ortler Group. Six of the ten highest peaks of Austria lie in the Ötztal Alps, including Austria's second highest mountain, the *Wildspitze* (3768 m), as well as *Weißkugel* (3738 m) and *Similaun* (3599 m).

Geologically, the Ötztal Alps are one of the classic locations of the Upper (formerly Middle) Austro-Alpine crystalline basement. The so called Ötztal Basement has experienced several metamorphic events: during the Variscan mountain building events around 300 Ma and again during the Eo-alpine event at about 80 – 90 Ma. Today, the rocks are marbles, garnet mica schists, amphibolite and orthogneiss. South of the Ötztal Basement there are the rocks of the so called Eo-alpine high pressure belt: the Schneebergerzug and Texel Group. The metamorphic evolution of these is very similar to the Kor- and Saualpe much further east. The Schneebergerzug is a 5 – 10 km wide band of strongly folded high grade metamorphic parageneisses between Sterzing and the Texel Group near Meran south of the Timmelsjoch pass. During Eo-alpine tectonism (80 – 90 Ma) the Ötztal Basement was folded together with the Schneebergerzug. Despite all this folding and metamorphism, there are several places near the

contact of the Schneebergerzug with the Ötztal Basement where there are Triassic carbonates that are still in their original sedimentary contact with the underlying crystalline rocks. For example the *Schneeberger Weißen, Moarer Weißen* (2825 m) and *Telfer Weißen* are made from dolomite that rests on crystalline basement. These island-like peaks were once connected with the Tribulaun and other peaks of the Brenner Mesozoic further east. At the contact between the dolomites and the underlying mica schists there are mineralisations of lead, zinc, copper and silver. The historic OK mines in the region of the Schneeberger Hut are considered the highest mines in Europe. They are located more than 2000 m above sea level.

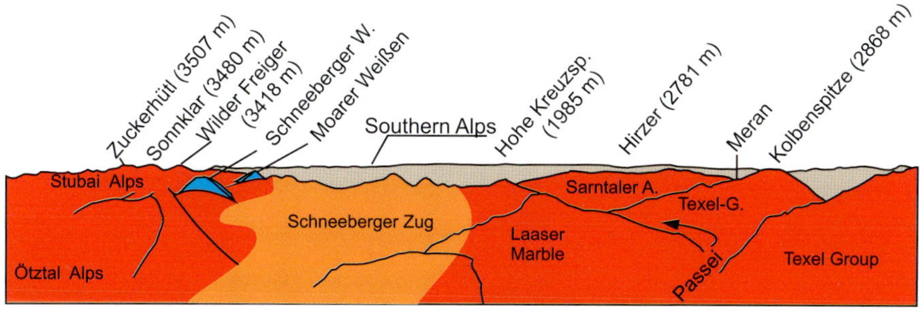

Right: Stammerspitze (3 254 m) is the 2nd highest peak of the Samnaun Group south of the Tyrolean city of Landeck (the highest is the Muttler, 3 294 m). The mountain is considered one of the least climbed mountains in the Eastern Alps. The summit region of the Stammerspitze is made up of layered carbonates, quartzites and radiolarites that belong to the Lower Austro-Alpine. This Lower Austro-Alpine klippe remained like an island on top of the enormous masses of Bündnerschiefer during the opening and erosion of the Engadine Window. The colours in the geological sketch correspond to those on the map below.

Below: Geological map of the Engadine Window. The red circle shows where the photo at right was taken.

Page 195: A view of the southeast margin of the Engadine Window looking southwest. In the foreground the settlement of Nauders, at right the Inn Valley. The valley at left leads to the Reschenpass at the border of Austria and Italy. The Reschenlake just behind it drains south towards the Vinschgau region. The boundary of the Engadine Window is formed by the tectonic Engadine Line. This line cannot be seen in the landscape. It runs through the mountains along the south side of the Inn Valley and leads all the way to the north side of the Bernina Group seen on the skyline.

The Engadine Window is one of three tectonic windows in the Eastern Alps in which the rocks of the former Penninic ocean floor may be seen to penetrate through the Austro-Alpine nappes. The other two windows are the Tauern Window and the Rechnitz Window (p. 74). In all three windows tectonics and erosion have removed many kilometre-thick layers of rocks of the Austro-Alpine nappe pile to free the view onto the rocks underneath.

The Engadine Window is about 50 km long and 10 km wide and its long axis lies along the upper Inn Valley between the townships of Prutz in the northeast and Tarasp southwest of Scuol. Compared to the other two windows, the Engadine Window is unique as it is the only window in which the en-

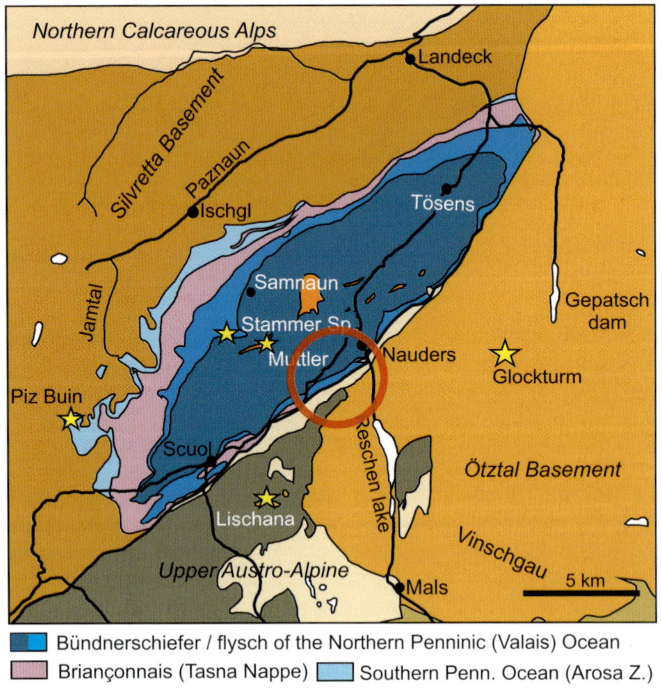

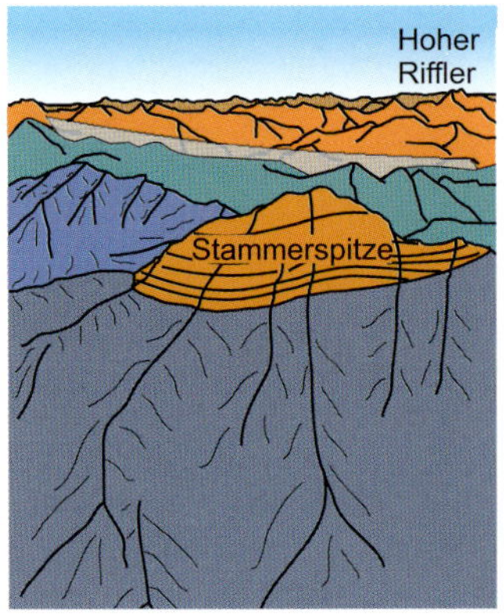

tire realm of the Penninic oceans may be recognised: much of the central window is made up of Bündnerschiefer and ophiolites of the north Penninic (Valais-) Ocean. The peaks of the Samnaun Group lie mostly in this central part of the window (Hohe Muttler 3 293 m, Piz Arina 2 828 m, Stammerspitz 3 254 m). The Tasna Nappe is made up of granites and gneisses that are covered by a thin veneer of sediments and form a thin hull of rocks around this central part of the window. It is considered to be part of the continental Briançonnais Fragment. Piz Tasna (3 179 m) north of Scuol forms the type locality for these rocks. Finally, the upper- and outermost part of the Engadine Window is made up of rocks that may be attributed to the South Pennininc Ocean: Bündnerschiefer and deep sea sediments that correspond to the Arosa Schuppenzone.

Along its southeastern side, the Engadine Window is bound by the Engadine Line; a huge structure that continues southwest of the window past the Bernina Massif and continues to the Bergell region. The fact that almost none of the boundaries of the window are along any of the major valleys indicates that the opening of the window already occured about 10 – 20 Ma and that the bounding structures were not active anymore during the landscape forming processes of the Alps.

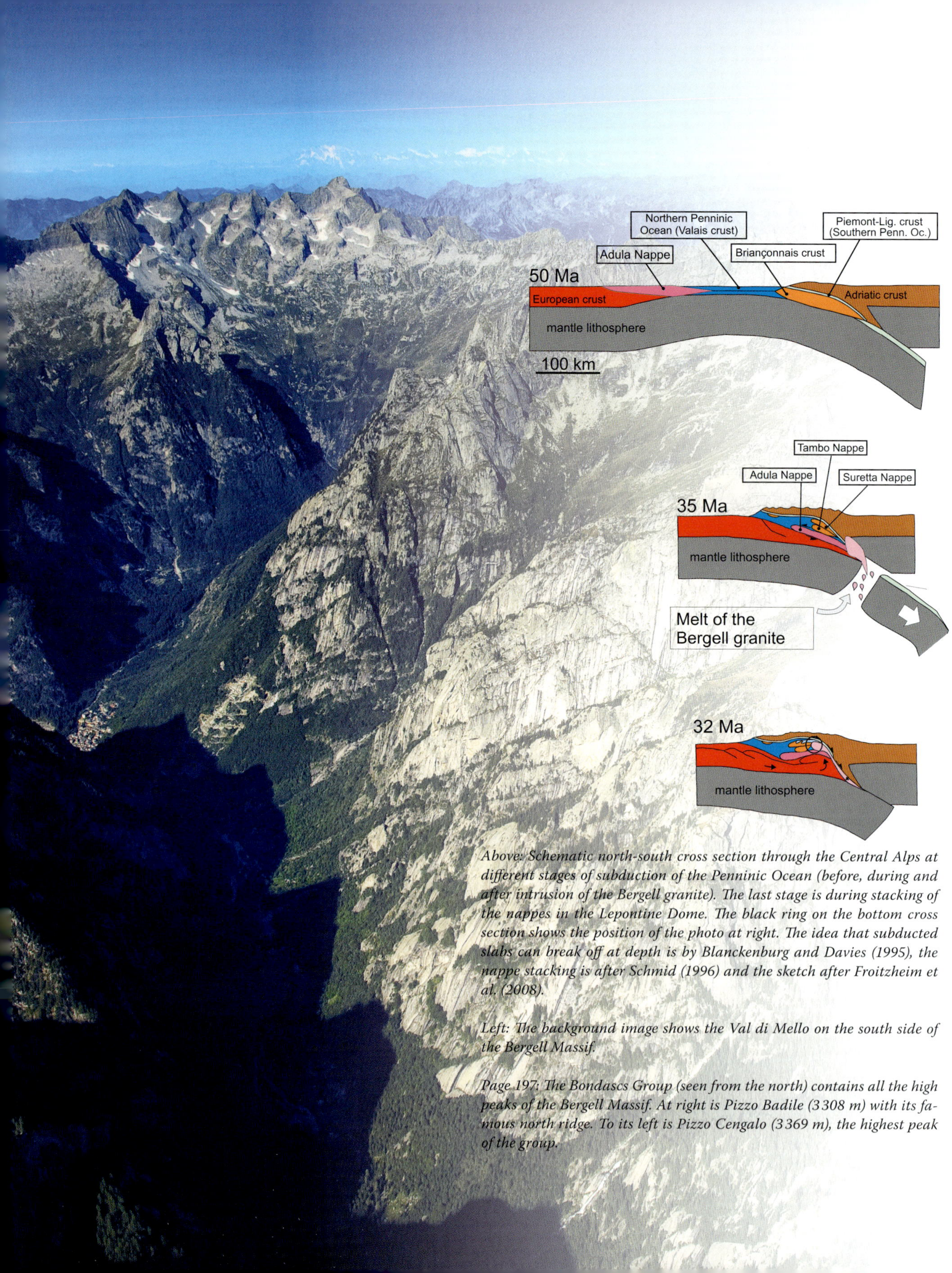

Above: Schematic north-south cross section through the Central Alps at different stages of subduction of the Penninic Ocean (before, during and after intrusion of the Bergell granite). The last stage is during stacking of the nappes in the Lepontine Dome. The black ring on the bottom cross section shows the position of the photo at right. The idea that subducted slabs can break off at depth is by Blanckenburg and Davies (1995), the nappe stacking is after Schmid (1996) and the sketch after Froitzheim et al. (2008).

Left: The background image shows the Val di Mello on the south side of the Bergell Massif.

Page 197: The Bondascs Group (seen from the north) contains all the high peaks of the Bergell Massif. At right is Pizzo Badile (3 308 m) with its famous north ridge. To its left is Pizzo Cengalo (3 369 m), the highest peak of the group.

Bergell is the region around the upper valley of the Mera, west of the Maloja Lakes near the Swiss-Italian border. For the geologist, the Bergell Massif is the second largest granite massif of the Alps (with the largest being Adamello Massif, p. 111). The granites of the Bergell Massif reach from the Mera Valley in the north to Monte Disgrazia in the east and to the valley of the Adda in the south. In the west, a long narrow band of granites continues almost to Lago Maggiore. The main peaks of the granite massif are Pizzo Badile (3 308 m), Pizzo Cengalo (3 369 m), Cima di Castello (3 388 m) and Monte del Forno (3 214 m). Val Masino and Val di Mello are two of the major valleys that lead deep into the massif from the south. The entire region is well known to rock climbers as the Alps as a whole offer few other granitic climbing regions.

The granites of the massif intruded the Bergell region about 35 Ma into the newly developed Periadriatic Line (p. 110). Chemically, the rocks classify as tonalite near the margins and as granodiorite in the central intrusion. Both rock types are very similar to granite. The granodiorite of the Bergell Massif contains many small dark inclusions that consist of material derived from the Earth's mantle. This observation has long lacked a good explanation, because geologists thought that – at the time of intrusion – the subducted Penninic Ocean lay between the Alpine crust and the Earth's mantle, thus shielding the crust from the involvement of mantle material. Today it is known that subducted plates can break off at depth. Some 35 Ma ago a tear formed in the subducted plate of the Penninic Ocean and it broke off. Magmas found their way through this tear and intruded into the middle crust forming the Bergell intrusion. Tectonic movements around this time deformed the margins of the intrusion and interleaved the granite with the surrounding rocks.

The Bernina Group lies along the Swiss-Italian border between the Upper Engadine region to the north and the Veltlin region to the south. The Bernina and Muretto Passes bound the massif to the east (towards the Ortler Group) and to the west (towards Bergell), respectively. The highest peak of the group is Piz Bernina (4049 m). It is the only mountain of the Eastern Alps that exceeds 4000 m in elevation. The other high peaks of the massif are (in the photo from left to right) Piz Palü (3900 m), Bellavista (3922 m), Piz Zupo (3996 m), Crast Agüzza (3854 m) and Piz Roseg (3937 m). The Bernina Massif is drained by the Morteratsch, Pers and Roseg glaciers. The *Morteratsch-Glacier* is considered to be the largest glacier of the Eastern Alps by volume, but it has lost more than 2 km in length over the last 100 years.

The geological significance of the Bernina Group is that it forms the only region of the Alps where the Lower Austro-Alpine rock units are visible over any major geographic extent. The Lower Austro-Alpine is the tectonically deepest unit of the Austro-Alpine nappe pile and is thus covered by the Upper Austro-Alpine rocks in most of the Eastern Alps. It is only visible near the rims of the Upper Austro-Alpine units: in the marginal regions of the Tauern Window and near the Rechnitz Window. In the Bernina region, Lower Austro-Alpine appears between the Penninic rocks of the Central Alps and the Upper Austro-Alpine of the Eastern Alps. The entire region, between Piz d'Ela and Piz d'Err to the north and the valley of the Adda (Veltlin) to the south, the Maloja Lakes to the east and Pontresina to the west including the Bernina Group in its centre, is included in this. Here, the Lower Austro-Alpine may be subdivided into two tectonic nappes: The higher Bernina Nappe and the lower Err Nappe.

In the Mesozoic, the Bernina Region was located along the northern margin of the Adriatic plate along its boundary to the Pennininc Ocean. Because of this, many of the Mesozoic sedimentary rocks are breccias made from submarine debris that once accumulated on the continental shelf. However, Piz Bernina itself is made up of diorite and gabbro and Piz Palu of granite, all of which are of Variscan age. In between those there are breccias and deep sea sediments. On the Morteratsch glacier there is a colourful mixture of all these rock types. Serpentinites and gabbros that lie underneath the Bernina Nappe formed during the opening of the Penninic Ocean.

Left: The "stairway to heaven of the Alps": The famous Biancograt ridge.

P. 199: The Bernina Group from the north. The Morteratsch glacier at its base is well known for its extremely rapid retreat.

Below: Geological west-east cross section through the Bernina Group. The colours correspond to those of the map and block diagram on p. 200. The location of the cross section is marked on p. 200 as a black line.

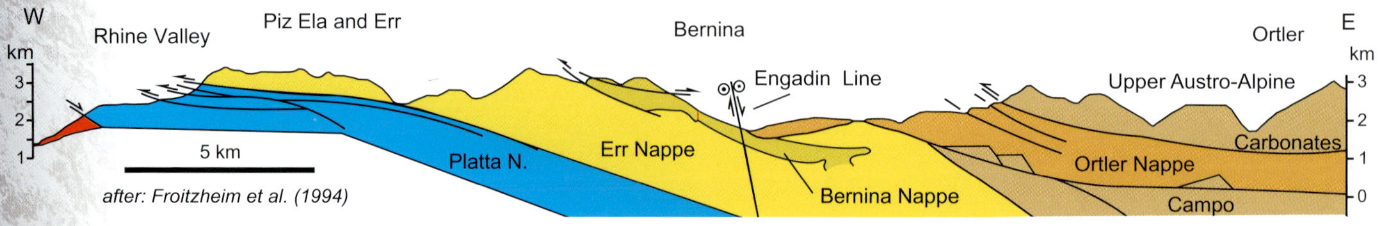

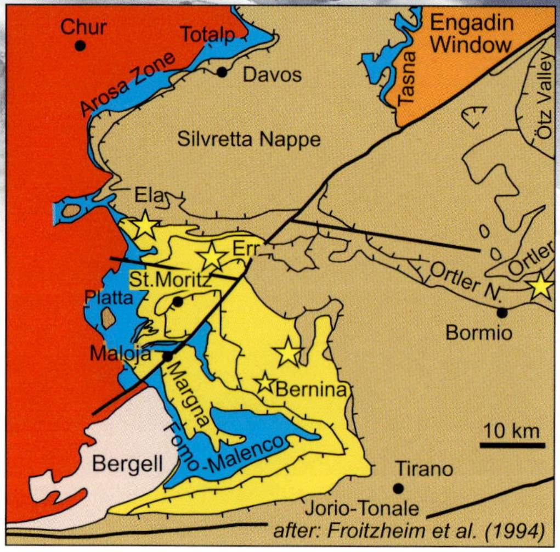

Above: The mountains Presanella (3556 m) and Adamello (3554 m) at left and Punta San Matteo (3678 m) at right.

Page 201: Ortler (3905 m), Cevedale (3769 m) and Zufallspitze (3752 m) (from left to right).

The Ortler (3905 m) is the highest peak of South Tyrol and its north face is considered the highest ice face in the Eastern Alps. Following the Bernina Massif, the Ortler is also the second highest mountain of the Eastern Alps. During World War 1, the highest military fighting in the Alps happened on the Ortler.

Geologically, much of the Ortler is made of Triassic Hauptdolomit, as is common in the Northern Calcareous Alps. However, tectonically the mountain belongs to a very different part of the Austro-Alpine: the *Central Alpine Mesozoic Cover*. This includes all the carbonate massifs between the Southern Alps and the Northern Calcareous Alps that are still in their original sedimentary contact with their crystalline base. The limestone mountains west of the Brenner Pass, the *Schneeberger Weißen* in the Passeier Valley, the *Stangalm region* in the *Gurktal Alps* or the *Lungauer Kalkspitzen* are part of this.

Ortler itself is made out of "Veltliner Basement" up to an elevation of about 2700 m. Similar to the Ötztal- or Silvretta Basement, the Veltliner Basement is made up of high grade metamorphic gneisses and schists with an ancient Variscan history. Above those, there is a thin layer of Permian volcanics and conglomerates (Alpine Verrucano) as well as lower Triassic sandstones and gypsum layers. Finally there is the 1000 m thick sequence of dolomite that makes up the upper parts of the peak. Together with the Eiger, the Ortler is one of the highest peaks of the Alps that is made up of sediments. The relatively high grade of metamorphism of the dolomite gives evidence for its tectonic history: during the Eo-alpine tectonic processes some 90 Ma ago, the Ortler was still buried deep in the Earth's crust underneath other Austro-Alpine nappes. Only much later was it exhumed by tectonic and erosive processes. Today, glaciers are receding rapidly and expose the crumbly dolomite.

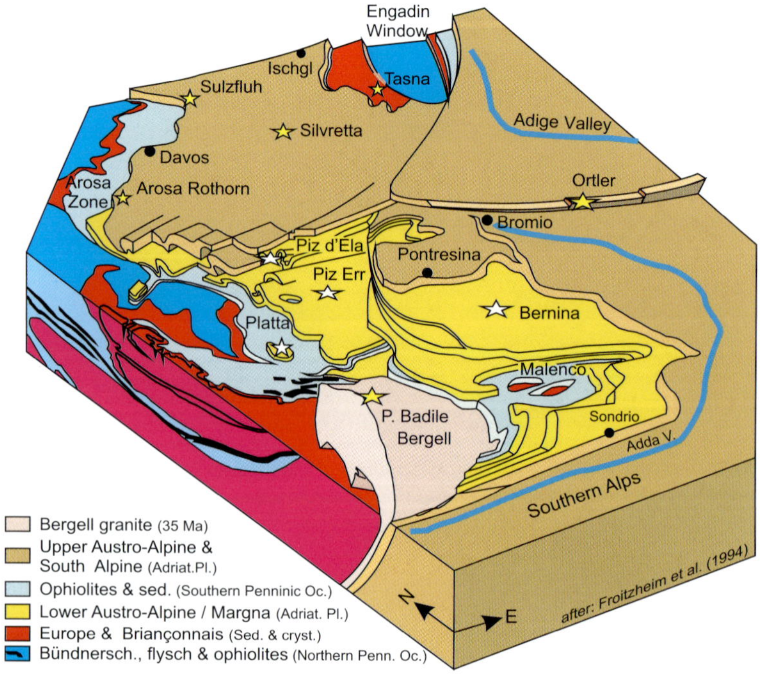

4.2. Southern Alps

The Rosengarten Group is part of the Dolomites (Southern Alps). View to east.

The Julian Alps are the secret sister to the famous Dolomites. They lie due east of the Dolomites in the Slovenian-Italian border region. They are bound by the valleys of the Save to the north, the Tagliamento to the west and the cities of Kranj and Ljubljana to the east. They are one of the easternmost parts of the Southern Alps and form the transition zone to the Dinarides. Geologically, the Julian Alps are part of the Southern Calcareous Alps that also include the Friuli Dolomites. As such, they consist of carbonates deposited in the Mesozoic along the southern margin of the Tethys Ocean. They correspond to the rocks of the Helvetic nappes that were deposited at the northern margin of the same ocean at the same time.

The dominating rock types are the 1000 m thick sequence of the Cordevol limestones (225 Ma) and the equally thick, clearly banked Dachstein limestone (200 – 215 Ma). Both units are separated by *Raibl beds* and overlain by a sequence of Jurassic rocks. Structurally, the history of the Julian Alps starts only in the Eocene (40 – 50 Ma), when the northern rim of the Adriatic plate was deformed by large thrust zones. Because of their position between the Adriatic plate, Dinarides and Alps, the plate tectonic stress field in the Julian Alps has changed several times over the last 30 Ma, so that faults in several orientations developed.

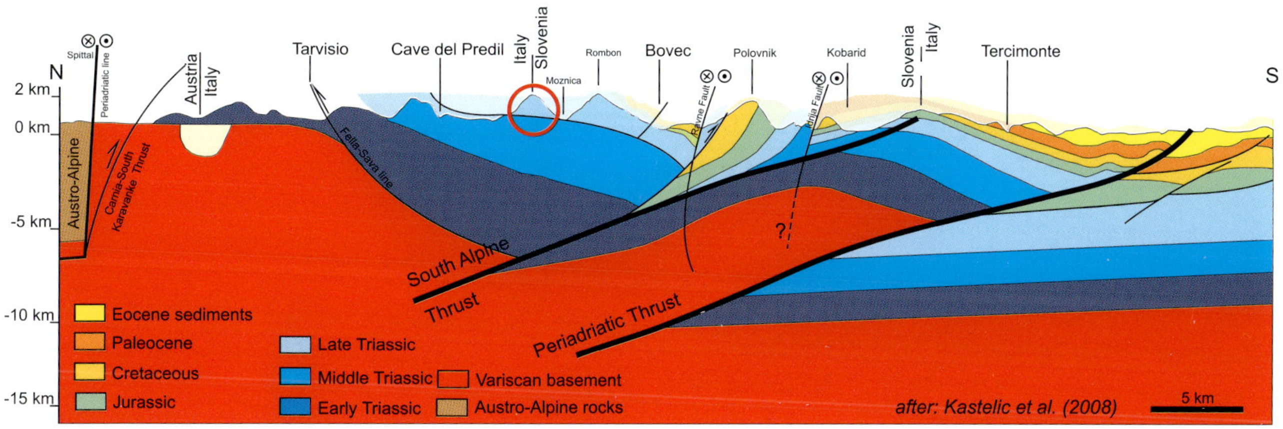

Above: The "Big Four" of the Julian Alps: Triglav (2864 m, highest peak of the Julian Alps and of Slovenia), Škrlatica (2740 m), Jalovec (2645 m) and Mangart (2679 m). The large rock face in the foreground is the Mojstrovka west face, in the middle of which the transition from the more massive Cordevol limestone to the banked Dachstein limestone is clearly visible.

Left: Geological north-south cross section through the Julian Alps (red ring = rough position of the photo above).

Page 207 and sketch on this page: The summit pyramid of Mangart in the Julian Alps displays a good part of the Mesozoic carbonate sequence of the Julian Alps in one little region. Jurassic rocks folded in the Paleogene (Plattformkalk, Sedlo-, Skrile- and Travnik formation) lie both above and below the massively banked Triassic Dachstein limestone.

Below: Jalovec and Mangart seen from high above Villach.

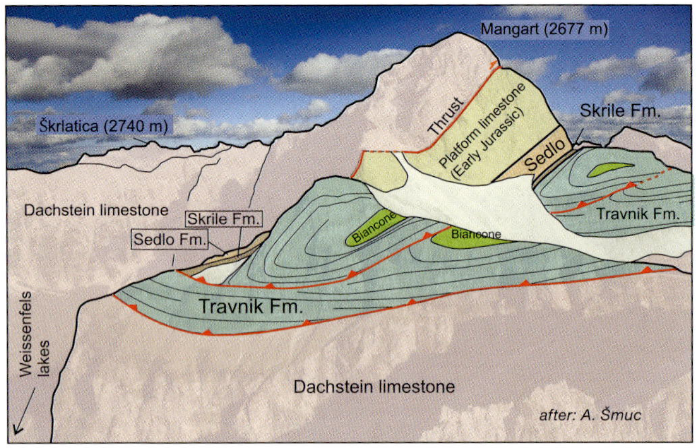

The Mangart of the Julian Alps: Mangart (2677 m) is the second highest peak of the Julian Alps and includes a spectacular exposure of all the rock types of the range in close spatial proximity. The base of the mountain is made up of a 1 km thick layer of Cordevol limestone. Above this, there is another kilometre of banked Dachstein limestone. Both were deposited in the upper Triassic (200 – 225 Ma) along the southern margin of the Tethys Ocean. On the Vršič Pass road the transition from the Cordevol to the Dachstein limestone can be seen.

At Mangart these rock units build up the dramatic north faces above the Weißenfelser Lakes. Above these, younger rock units follow, telling a fascinating story of the Tethys Ocean in the Jurassic. Initially, the ocean floor sank down. Reef platforms developed on top of the Dachstein limestone. These limestones contain hornfels nodules and spherical carbonate concretions. The ocean floor sank more rapidly than the reefs could grow, so the reefs drowned. Mud rich limestones formed that contain ammonite and sponge fossils (*Sedlo Formation*, 185 Ma). Above this follows the *Skrile Formation* (180 Ma), made from deep water carbonates containing radiolarites. Above that follows the *Travnik Formation* (150 – 170 Ma), made from a thinly layered sequence of schists, radiolarites and limestones. This unit indicates that the water was still deep but that coral reefs were not far in shallow water. Geologists speak of the deep "Bovec Basin" and the shallow "Julian High" to describe these Jurassic deep and shallow water basins. During Paleogene and Neogene tectonics, the soft Jurassic rocks were folded and the massive Dachstein limestone thrust over them. As a consequence the Jurassic sequence is partly upside down on Mangart.

Above: *The Slovenian city of Bled and the southern side of the Karavanke Range. The highest peak of the Karavanke Range is seen near the left margin of the photo: Hochstuhl (2 237 m). The peak in the sun in the photo centre is called Koschuta (2 136 m). Behind its right shoulder the Grintavec Massif (2 558 m). The sudden jump in the topography at the base of the range is defined by the right lateral Sava fault. This fault has 30 km of displacement.*

Right: *Geological map of the Julian Alps, Karavankes and Steiner Alps. The red ring shows the position of the photo above.*

Page 209: *The rocks around the Wolayer Lake (on the mountain: Hohe Warte, 2 780 m) in the Carnian Alps display a unique juxtaposition of different Paleozoic carbonates. For this reason, UNESCO lists the region as part of the hundred most important geological sites in the world (see also p. 14).*

Karavanke and Carnic Alps: The Karavanke Range and the Carnian Alps are two long mountain ranges along the border between Austria, Italy and Slovenia. They also form the boundary between Eastern- and Southern Alps.

The *Karavanke Range* consists largely of Triassic carbonates and is geologically very similar to the Julian Alps and the Dolomites. However, the Periadriatic Line (p. 110) cuts the range lengthwise into two parts: the northern and the southern Karavankes. Between them, there lies a narrow band of older Paleozoic crystalline rocks. This band of rocks was dragged into its position amidst the carbonates by many kilometres of displacement along the Periadriatic Line. This band of crystalline rocks leads through the township of Eisenkappel and includes a granitic body that belongs to the 30 Ma old series of intrusives along this line (see p. 110): the *Karavanke Granite.*

The Slovenian township of Bled on the south side of the Karavankes lies in a topographic depression that was carved during the glaciation periods by the Bohinjer Glacier that then filled the Save Valley. The terminal moraine of this glacier may be seen between Bled and Kranj. When the glacier retreated, the melting ice formed Lake Bled. For this reason, the lake has neither inflow nor outflow. In the photo at left, the river Save flows from left to right (west to east) just 1 km behind the lake.

The Carnic Alps are the next range to the west of the Karavankes. They are mainly made up of Paleozoic carbonates of Silurian, Devonian and Carboniferous age (p. 94). The limestones are significantly older than those of the Karavankes or those of the Northern Calcareous Alps. They are more comparable to the rocks of the Highlands of Graz or the Greywacke Zone. On the mountains Kellerwand and Hohe Warte (2 780 m), high above the Wolayer Lake, these ancient coral reefs are up to 1 300 m thick. Tectonic motions placed these reefs directly opposite deep sea sediments. The Carnic Alps are famous for this unique juxtaposition, and for this reason, are considered part of the hundred most important geological sites in the world.

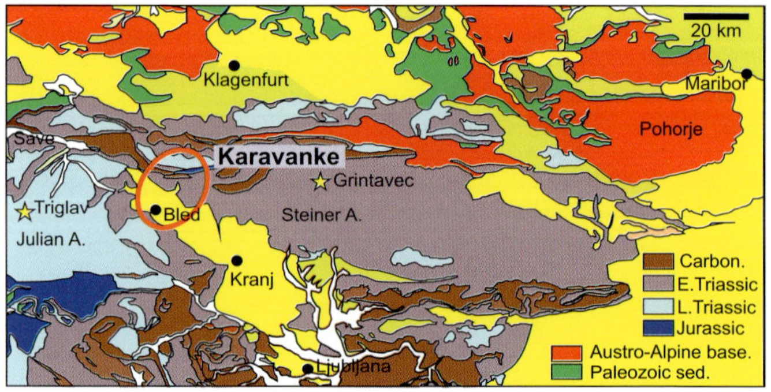

Above: The southern rim of the Alps in the Piave Valley north of Belluno in northern Italy.

Right: Geological map of the southernmost part of the Alps (Ring = photo, p. 211).

Page 211: The Po plains in the region of the river Piave in the Italian province Venetia. On the skyline the Adriatic Sea. The town in the photo centre is Bassano del Grappa. The arc-shaped chains of hills are made up of sandstones of the Molasse Zone. They are thrown into wave-like hills by the tectonic southward push of the Alps.

The Po plains bound the Alps to the south. This huge and almost flat region around the river Po corresponds largely to the geological region of the southern Molasse Zone. Just as the northern Molasse Zone, the southern Molasse Zone is a classic foreland basin. Foreland basins are sedimentary basins that are formed by the load of a rising mountain belt that bends a plate downward into the Earth's mantle – similar to the way an air mattress in water is bent down by a load. In the Molasse basins, this bend was filled by sediments at the same rate as it formed. Therefore, the surface of the foreland basins always remained around sea level and is completely flat. The oldest sediments at the base of the kilometre thick pile that fills the Molasse basins are around 35 Ma old. They were deposited when the mountain range adjacent to it began to rise.

The rivers Tagliamento, Piave, Adige and Aosta are among the most important rivers that drain the Alps to the south and deposit their sedimentary load into the southern Molasse Basin.

Two geologically very different arc-shaped chains of hills may be found near the transition from the Alps into the Po plains. One set of hills are moraine walls from the last glaciation period. These hills are always near the large rivers and along the southern rim of the large Italian lakes. The other arc-shaped set of hills are folded Molasse sediments that are thrown into wave-like structures. These hills are signs that the Alps slowly push tectonically southwards and successively incorporate the Molasse Zone sediments into the mountain building processes.

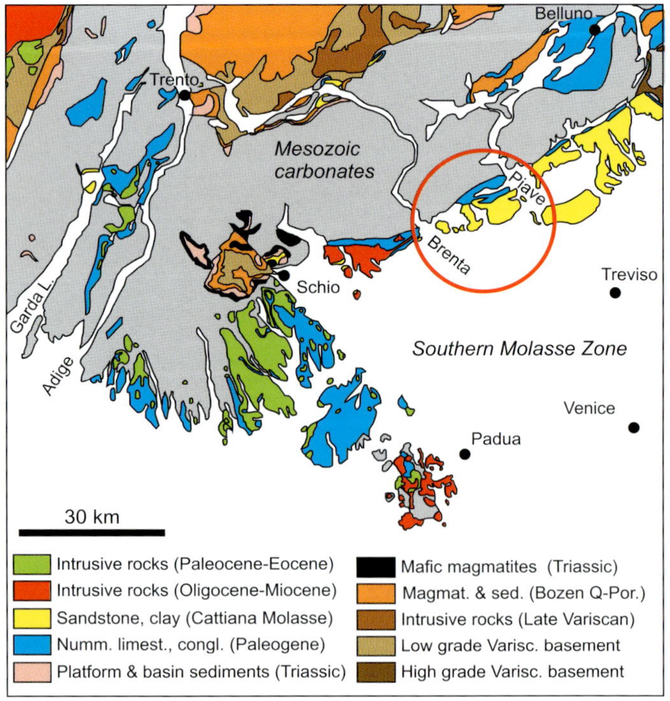

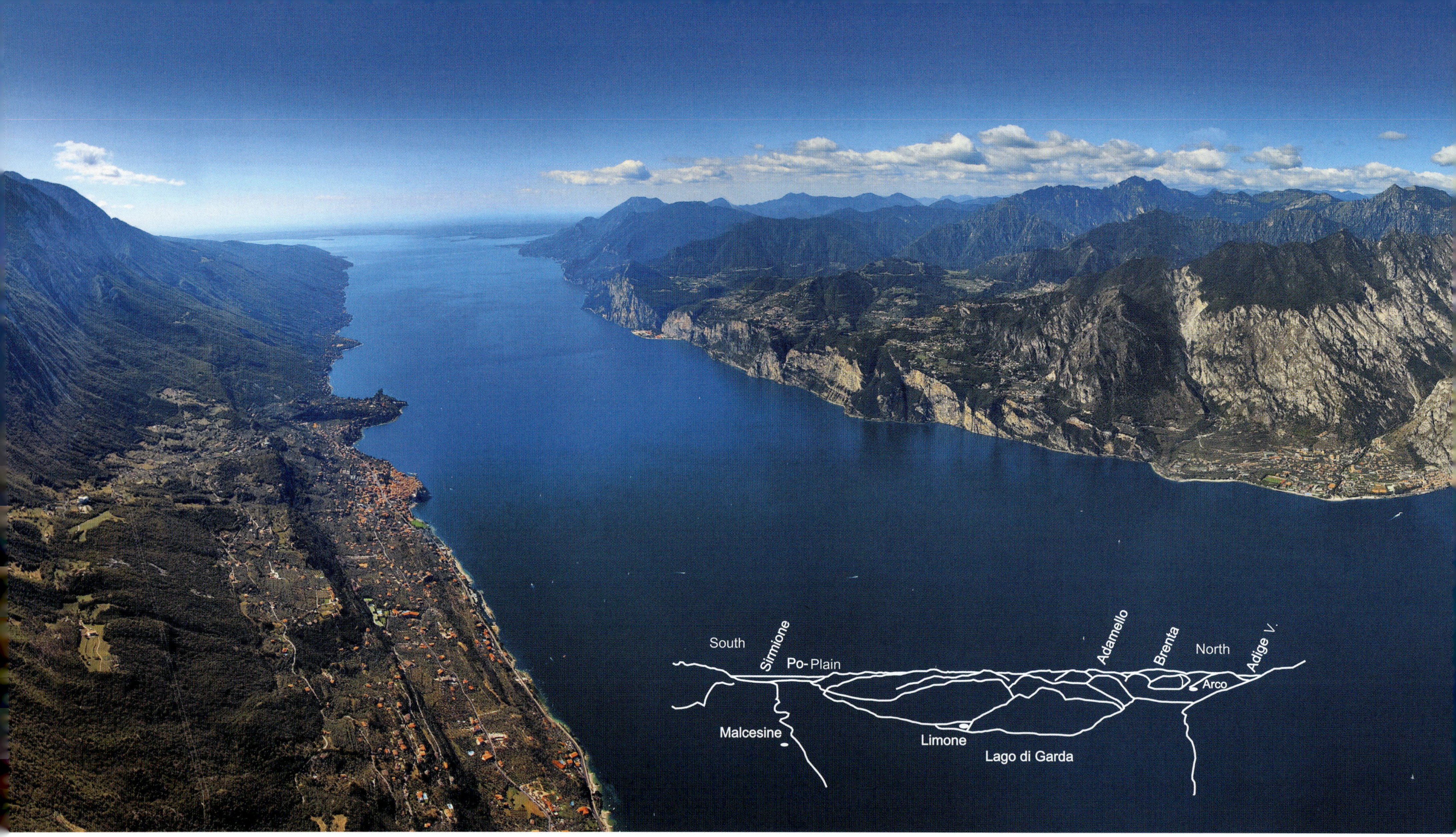

Above: Lago di Garda is the largest of the Italian Lakes along the southern margin of the Alps. The lake and its surrounding mountains are a famous playground for climbers, mountain bikers, paragliders and wirdsurfers. The valley of the river was carved five million years ago when the sea level in the Mediterranean was about 1 kilometre lower than today.

The Large Italian Lakes: Along the southern margin of the Alps there is a series of large lakes that have a very similar geology and geometry: from west to east these are: *Lago Maggiore* (212 km^2), *Lago di Lugano* (50 km^2), *Lago di Como* (146 km^2), *Lago d'Iseo* (65 km^2) and *Lago di Garda* (370 km^2) as well as a series of smaller lakes, for example *Lago d'Orta*, *Lago di Varese* and *Lago d'Idro*. All have in common that they are north-south oriented, with one end being located deep inside the Alps and the other in the Po plains. The most important feature is that the floor of most of these lakes lies significantly below sea level. In fact, in Lago Maggiore, Como and Garda it is even lower than much of the Adriatic sea floor today. Thus, these lakes cannot have been carved during the ice ages, as glaciers float on water when their valleys reach sea level. It must be concluded that the valleys of the lakes were carved at a time when the Adriatic sea level was significantly lower than today: during the Messinian crisis. At about 5.9 Ma the straits of Gibraltar were closed, the Mediterranean evaporated and for about 500000 years, the sea level was about 1 km lower than today. In this time the south flowing rivers carved

their valleys deep into the Alps and into the Molasse sediments in the foreland. In the Miocene, canyons hundreds of metres deep, were carved into the kilometre thick sediments of the Molasse Zone (p. 215). When the straits of Gibraltar reopened half a million years later and the Mediterranean sea was re-flooded, these canyons were quickly refilled by the sediments of the Po damming up the lakes behind them. Today, these refilled canyons can be seen with geophysical methods.

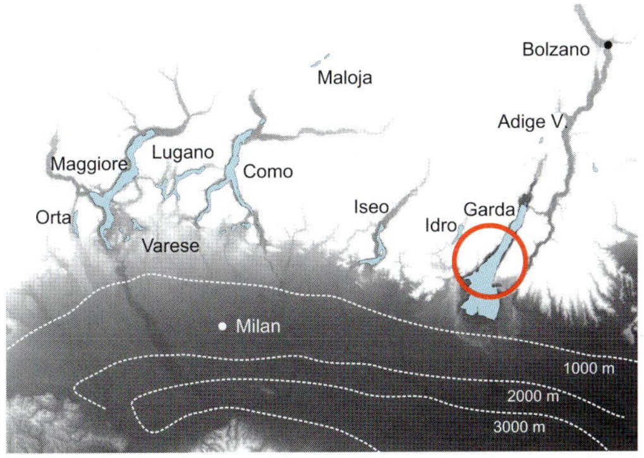

Left: Map of the large lakes along the southern margin of the Alps. The transition from white to black is the topographic boundary of the Alps and also the geological boundary between hard rocks and sediments of the Molasse Zone. The contours show the thickness of the sediments in the Molasse Zone (ring = photo above).

Above: The island in Lago d'Iseo is the largest in any freshwater lake in Europe: Monte Isola. The two smaller islands to the left and right behind it are Isola di San Paolo and Isola di Loreto. The river Oglio flows through the lake and cuts through the glacial moraine landscape in the background. These moraines were formed in the last ice age when the lake was filled by glaciers.

Lago d'Iseo is one of the smaller lakes of the large north Italian lakes, but it contains the largest island in any body of fresh water in Europe. As for the valleys of the other lakes, the valley of Lago d'Iseo was formed about 5 Ma. The river that flowed along it ran through a deep gorge that cut hundreds of metres below today's surface into the Molasse Zone sediments at the southern end of the lake. Only 2 million years ago, this canyon was refilled with river sediments and the whole region uplifted to its present day elevation above sea level. The filled canyons are well known from drill holes and geophysical measurements. During the ice ages, the lake was re-carved, its steep flanks were formed and the moraines at its southern end piled up. Today's drain of the lake cuts through these moraine walls.

Lake	Elevation of Lake Surface	Maximum Water Depth
Maggiore	193 m	372 m
Lugano	271 m	288 m
Como	199 m	410 m
Iseo	185 m	251 m
Garda	65 m	346 m

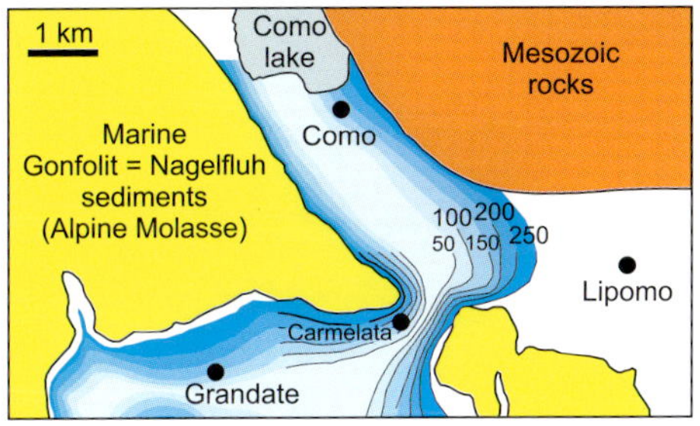

Left: The southern end of Lago di Como. The white and the blue region are filled today by glacial sediments. The labelled contours (in metres above sea level) show the depth of the sub-surface canyons through which the lake drained towards the Adriatic sea about 5 Ma (after: A. Bini, 1978).

The Dolomites are possibly the most famous part of the Southern Alps. Geographically, the range is bound to the west by the valleys of the rivers Etsch (Adige) and Eisack (Valle Isarco) and to the east by the Julian Alps. To the north, the Dolomites are bound by the Periadriactic Line along the Val Pusteria (Puster Valley). To the south the range reaches roughly as far as the valley of Belluno. Geologically, the Dolomites are made up of a thick sedimentary pile that was deposited between the Permian (about 240 Ma) and the Cretaceous (around 140 Ma). Most of the major massifs of the range consist of sediments that were deposited into shallow water of the Palaeotethys within the first 10 – 30 million years of this prolonged period – roughly at the same time as the Northern Calcareous Alps were deposited much further north. During Alpine tectonics, this sedimentary pile was barely affected and the sedimentary succession of the Dolomites is therefore largely preserved in thickness and geometry as it was deposited in the Mesozoic. As such, the Dolomites allow an unparalleled view of the sedimentary evolution all across the Mesozoic.

Famous names of the individual massifs like the *Sella Group, Marmolata, Pala Group, Rosengarten* or *Langkofel* are well known among climbers. The subdivision of the Dolomites into these spatially well-confined massifs (or: "groups") has a geological reason: The carbonates of the Dolomites form a roughly two kilometre-thick slab that lies on top of schists and volcanic rocks. Erosion has dissected this slab all the way down into the volcanic quartz porphyries and crystalline schist leaving behind the individual massifs like isolated islands.

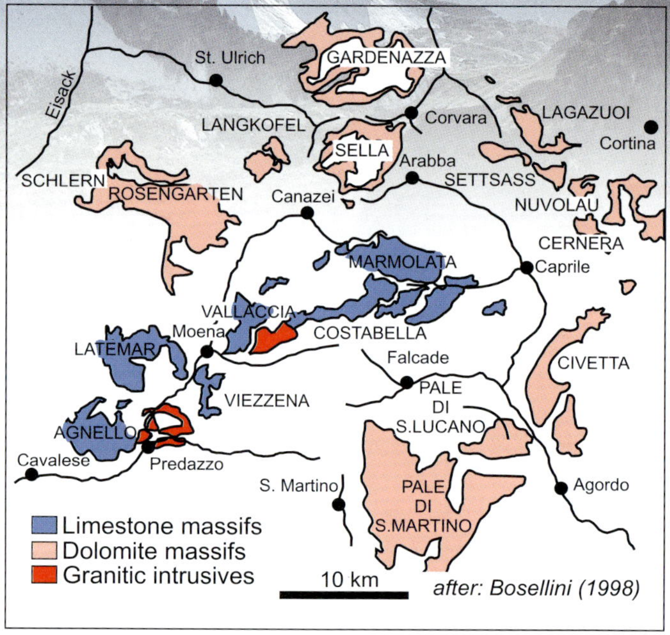

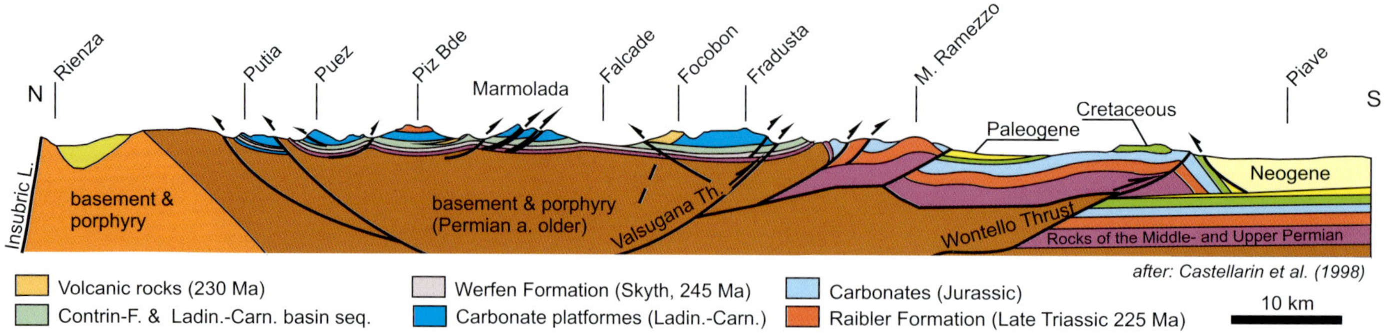

Right: Simplified geological map of the Dolomites.

Below: Geological north-south cross section through the Dolomites.

Page 217: The famous north faces of the Drei Zinnen (ital.: Tre Cime di Lavaredo). From left to right the summits are: Preußturm, Punta di Frida (2 792 m), Kleine Zinne (2 857 m), Große Zinne (2 999 m) and Westliche Zinne (2 973 m). The entire massif is made up of dolomite (Hauptdolomit), deposited between 215 and 200 Ma. The horizontal banking of the dolomite is caused by thin mud banks that interlayer the dolomite. The rhythmic alteration between dolomite and mud was created by small but regular sea level fluctuations caused by climatic fluctuations that – in turn – may have been caused by orbital changes of Earth around the sun.

The development of the Dolomites started in the Permian. Volcanic lavas erupted from an enormous caldera near Bolzano and covered more than 2 000 km² about 2 kilometres thick. These lavas are called the *Bozen Quarzporphyry* and are a popular building stone. Soon after the continent sagged and the volcanoes were covered by fluvial sediments (Grödener sandstone). When the surface reached sea level, the water of the Palaeotethys (Meliata-Hallstatt Ocean) flowed in. Salt and gypsum were deposited in the shallow lagoons and today form the Bellerophon beds. At the beginning of the Triassic, schists of the Werfen Formation were deposited over the evaporites. About 245 Ma ago, coral reefs started growing in the deepening water. These reefs build up the *Rosengarten Group, Marmolata, Pala Group* and many other massifs.

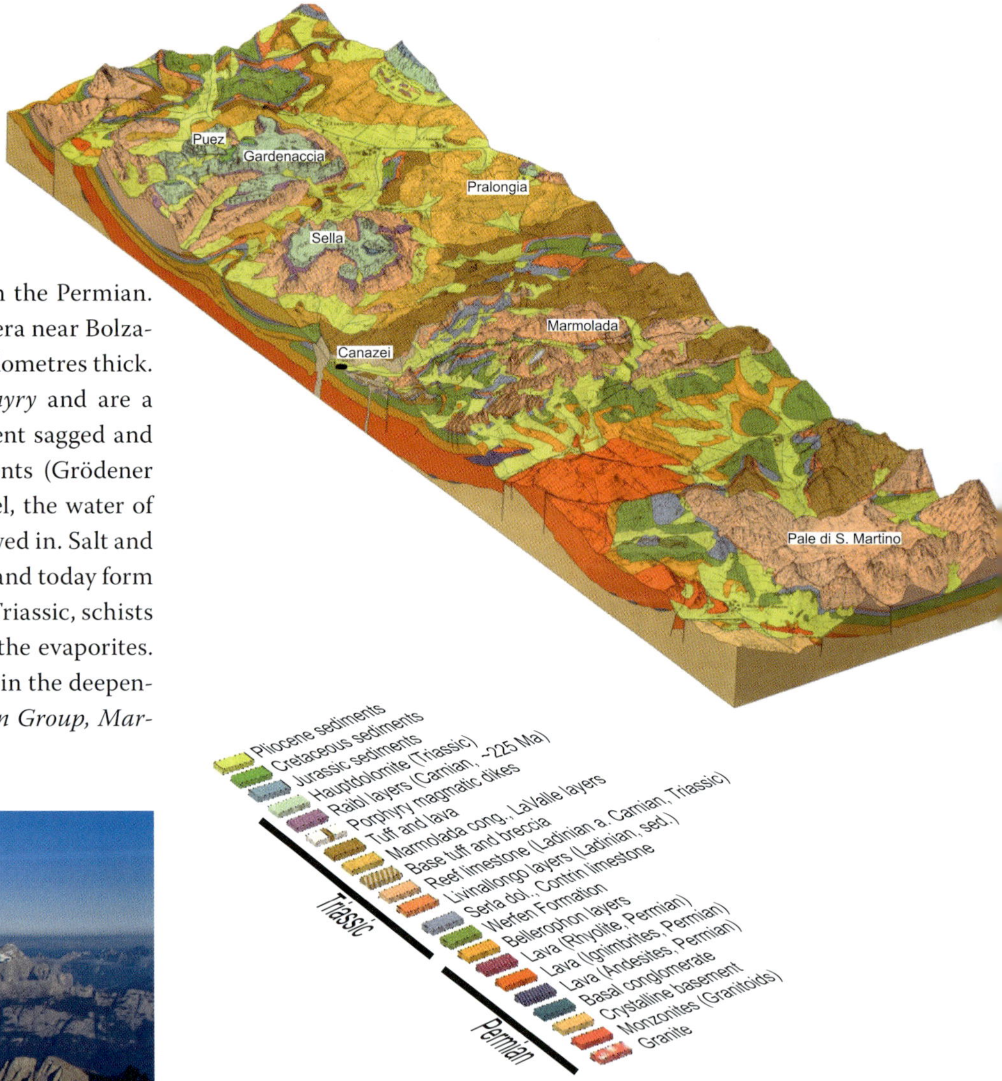

Above: Block diagram of the Dolomites. It may be seen that the carbonate massifs all lie on a crystalline base (after: P. Leonardi and A. Bosellini).

Below: The Marmolata Massif in the Dolomites from the west. Behind it on the right is Civetta (3 220 m) and – half covered – Monte Pelmo (3 168 m). On the very left the valley of Cortina d'Ampezzo.

Page 219: The Rosengarten Group with its high point the Rosengartenspitze (2 981 m) from the south. Behind the massif lies the valley of the Eisack between Bolzano (left) and Brixen (right). The Sarntal Alps with their highest peak the Hirzer (2 781 m) can be seen on the skyline. In the far background at right the Zillertal Alps. In the foreground it may be seen that the dolomites of the Rosengarten Group lie on a layer of red-coloured Werfen Formation rocks. The spectacular vertical face in the left foreground is the Rotwand (2 806 m). The long horizontal band of cliffs at the base of the massif is made from dolomite of the Contrin Formation. The Schlern dolomite above it shows the typical inclined layering of a southeast directed reef slope that migrates towards the open sea.

The Dolomites from the Mesozoic to the Present: Between 240 and 210 Ma huge carbonate platforms developed in the Dolomites. During this 30 million year long period, minor tectonic motion caused repeated uplift, tilting or submergence so that the growth of the coral reefs was repeatedly interrupted by erosion. These erosional surfaces can be easily recognised because of their irregularity and the discontinuous deposition of younger sediments.

At about 230 Ma, the reef development was interrupted by a volcanic event. In the region of the western Dolomites volcanoes erupted and the lava buried some of the coral reefs (e.g. Marmolata, Latemar and Agnello Massif). After this event the carbonate growth continued. As one of the last thick carbonate depositions, the Hauptdolomit was deposited about 205 Ma ago. This Hauptdolomit is equivalent to that of the Northern Calcareous Alps. However, Dachstein limestone is not very widespread in the Dolomites. In the Jurassic and Cretaceous, sedimentation occurred only sporadically and the Dolomites remained tectonically quiet. New tectonic motion initially started 40 – 60 Ma, pushing the Dolomites to the west. From about 20 Ma onwards the Dolomites were then thrust southwards along huge shallow fault zones like the *Valsugana* and *Wontello Thrust.*

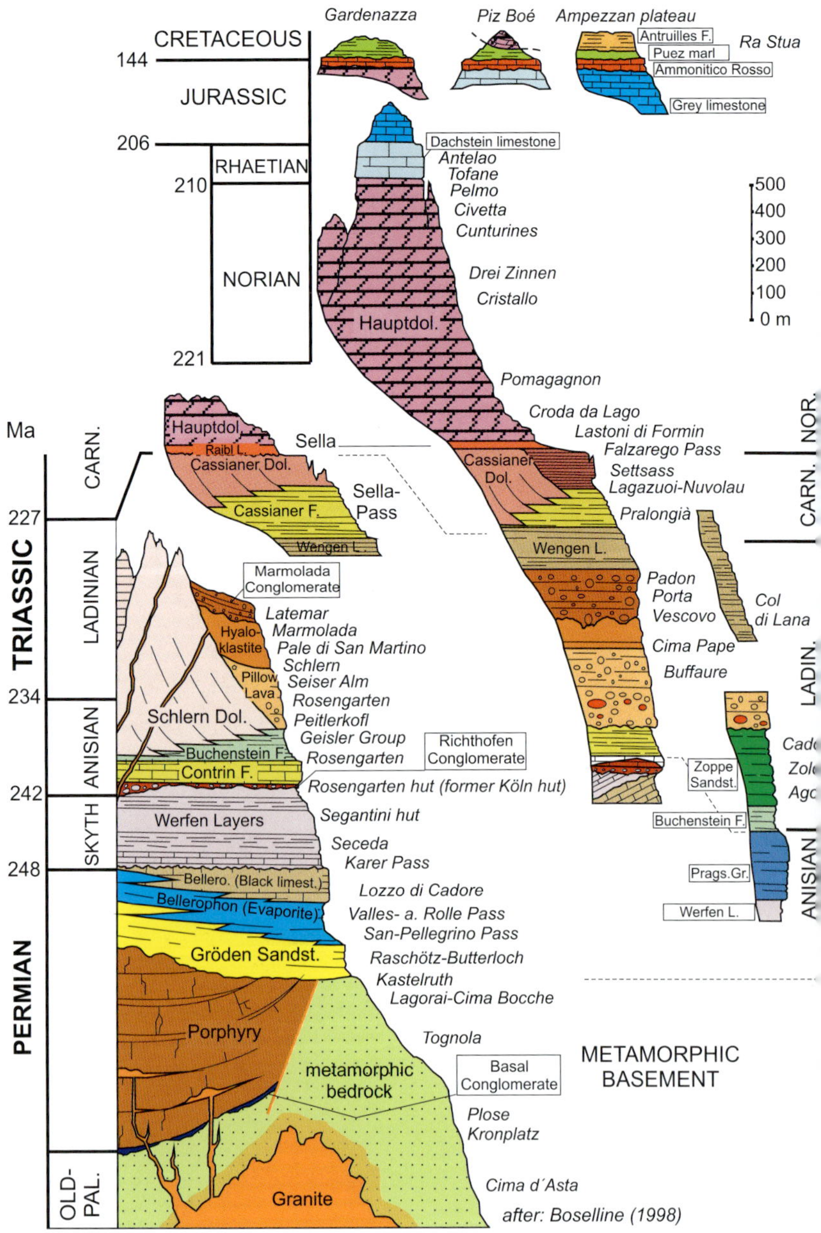

Left: The Schleierkante on the Cima della Madonna (2 733 m) at the south end of the Pala Group is predominantly made up of Schlern Dolomite and is one of the most famous climbing routes in the Alps.

Right: Schematic summary of the geological units of the Dolomites.

Page 221: The eastern part of the Pala Group with Monte Agner. The characteristic north ridge that separates light and shade on the photo is considered to be the longest climbing route in the Dolomites and one of the longest in the world.

Mt. Shiara
Mt. Agner (2872 m)
Sass del Caure (2762 m)
Croda Granda (2849 m)
Po Plain
Agordo
Mt. San
Lucano
Valle di
San Lucano
Altipiano
delle Pale di San Martino
Cima Vezzana
(3192 m)

4.3. Central Alps

The Valais Mountains are part of the Central Alps. View to South.

The Matterhorn (4 477 m) is probably the most famous mountain in the Alps. Although there are at least 10 higher peaks in its immediate surroundings, it is considered the jewel in the crown of the Valais Alps and stands high above the township of Zermatt at the end of the Matter Valley. In total, 43 of the roughly 80 peaks in the Alps that exceed 4 000 m elevation stand here in the Valais Alps. Geologically, the Matterhorn is part of the Dent Blanche Nappe: a 20 x 50 km sized klippe of 1 km thickness made up of crystalline gneisses and schists and are considered to be part of the Lower Austro-Alpine. As such, the Matterhorn is part of the Mesozoic Adriatic (Apulian) Plate. The Dent Blanche Nappe forms a free floating slab of the Adriatic plate that lies on top of the Penninc rocks of the Combin Zone. The rocks of the Dent Blanche Nappe also build up the mountains Dent Blanche itself (4 356 m), the Weisshorn (4 505 m), Zinalrothorn (4 221 m), Gabelhorn, Matterhorn and Mont Collon (3 637 m). To the southeast, the Dent Blanche Nappe reaches as far as the Aosta Valley.

Below the Dent Blanche Nappe lie the rocks of the south Penninic (Piemont-Ligurian-) Ocean floor which may be tectonically subdivided into the Combin Zone and the Zermatt Saas Zone. Both zones consist of Bündnerschiefer with ophiolites in between. North of Zermatt lie crystalline rocks of the Mischabel Nappe of the Briançonnais Fragment (p. 70). This nappe forms a huge fold in the Zermatt Valley called the Mischabel backfold. This fold pushes the rocks of the Mischabel Nappe over the top of the Combin Zone (p. 72).

Above: Geological sketch of the photo at right.

Right: Panorama sketch of the Valais Alps. All rocks below the Matterhorn once belonged to the South Penninic ocean floor.

Page 225: The Matterhorn is probably the best known peak in the entire range. The mountain is made up of gneisses that are part of an isolated slab of continentally derived rocks that lies as isolated klippe on top of the rocks of the South Penninic Ocean floor (Combin Zone). Tectonically, the rocks are part of the Dent Blanche Nappe that is considered to belong to the Austro-Alpine.

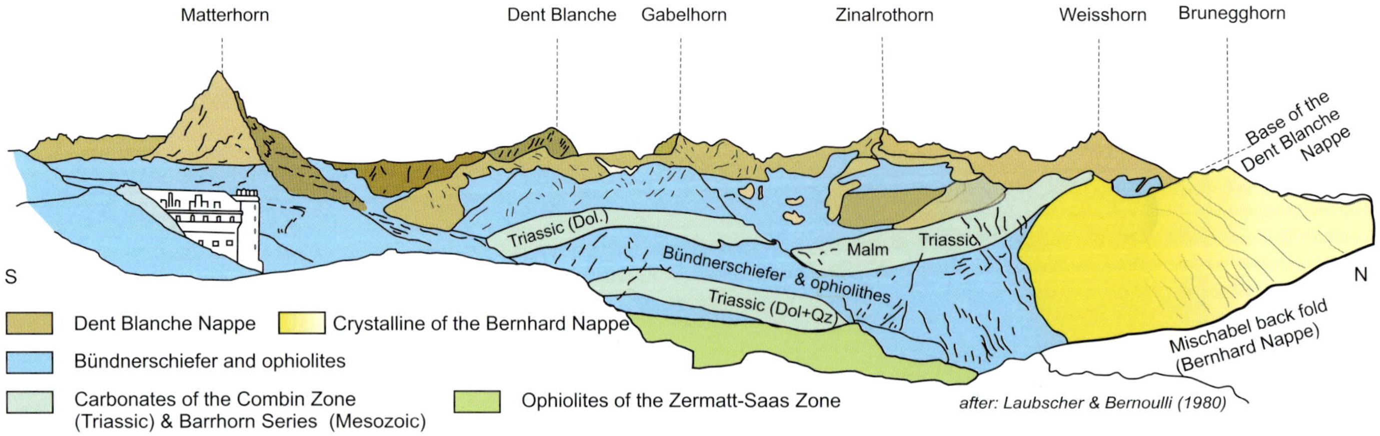

Mt. Blanc (4808 m) is the highest mountain of the Alps. 27 of the 82 peaks in the Alps that exceed 4000 m in elevation are within this massif. Geologically, the Mt. Blanc Massif is (together with its northern neighbour, the Aiguilles Rouges Massif) one of the five External Massifs of the Alps that formed around 300 Ma in the Variscan mountain building cycle. During the much later Alpine events it was barely deformed. The light coloured Mt. Blanc granite that is so well known to climbers intruded the European crust at 304 Ma, and the deformation of the surrounding rocks is also of about this age. During the Mesozoic, the Mt. Blanc and Aiguilles Rouges Massifs were buried under many kilometre thick-sediment layers. These sediments still mantle both massifs and a narrow band of these sediments separates the two massifs along the valley of Chamonix. The Alpine evolution of the Mt. Blanc Massif started with its uplift towards the surface about 22 Ma when the folding in the Helvetic Nappes began.

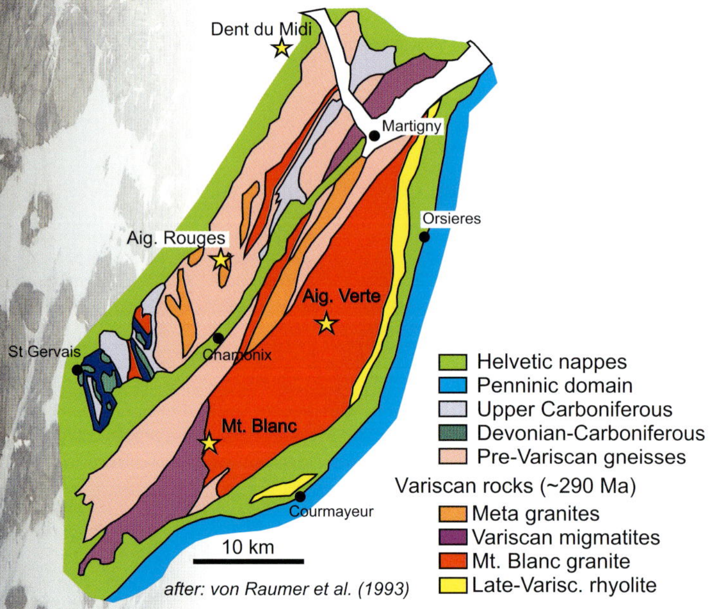

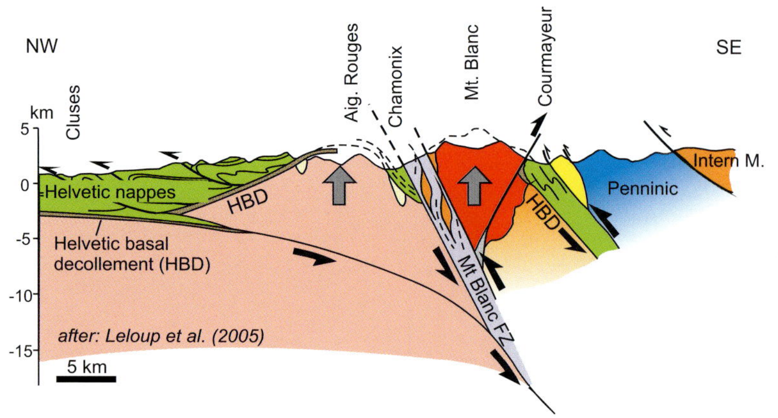

Because of this uplift, the kilometre-thick sediments on top of Mt. Blanc were domed up and rapidly began to erode. The unloaded massif started rising to the surface. At about 12 Ma this rise began to exceed that of the northerly Aiguilles Rouges Massif, and the Mt. Blanc Massif was moved somewhat over the top of the Aiguilles Rouges Massif along a thrust zone located in the Chamonix Valley. Even after all the sediments were eroded off the top of Mt. Blanc, this thrusting continued, and erosion exhumed the Mt. Blanc granite from the surrounding Variscan gneisses. About 4 Ma, the relative motion of the two massifs stopped. A fault zone south of Mt. Blanc became active and uplifted the mountain to its present day elevation. The unusually high elevation of the mountain has two reasons: firstly, two V-shaped fault zones squeeze the mountain upwards and secondly it is also due to the hard Variscan rocks it is made out of.

Above: Geological profile through the Mt. Blanc-Aiguilles Rouges Massif.

Photo at left: Walker pillar on the Grandes Jorasses is one of the classic north faces of the Alps.

Map at left: Geological map of the Mt. Blanc Massif.

Page 227: the south side of Mt. Blanc summit build up shows the light coloured Mt. Blanc granite. The Brouillard ridge to the left, Peuterey ridge to the right. Below and to the left of the summit the three Frêney pillars.

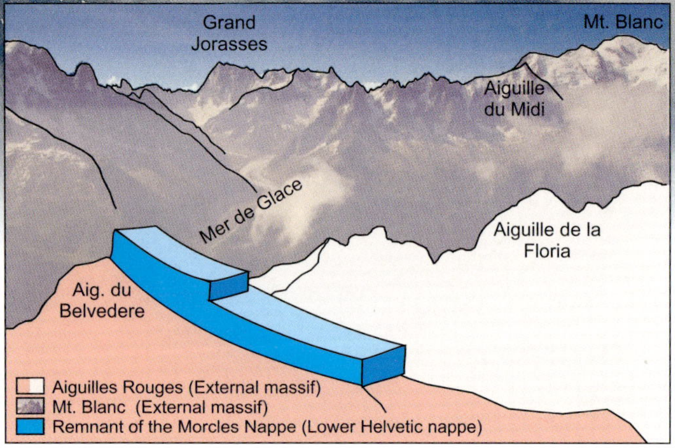

Above: Mt. Blanc seen from the Aiguilles Rouges Massif.

Right: Geological sketch of the photo on p. 229.

Below: Geological cross section through the Central Alps.

Page 229: The high peaks of the Aiguilles Rouges Massif in front of the skyline of Mt. Blanc.

The Aiguilles Rouges Massif is the northwestern neighbour of the Mt. Blanc Massif. Together the two massifs form one of the five so-called External Massifs of the Alps (p. 55). The Aiguilles Rouges Massif is about 50 km long and 10 km wide and centred around the Aiguilles Rouges (2966 m) north of Chamonix. To the southeast the massif is bound topographically and geologically by the valley of Chamonix. In the northwest it reaches to the lakes Lac de Salanfe and Lac d'Emosson.

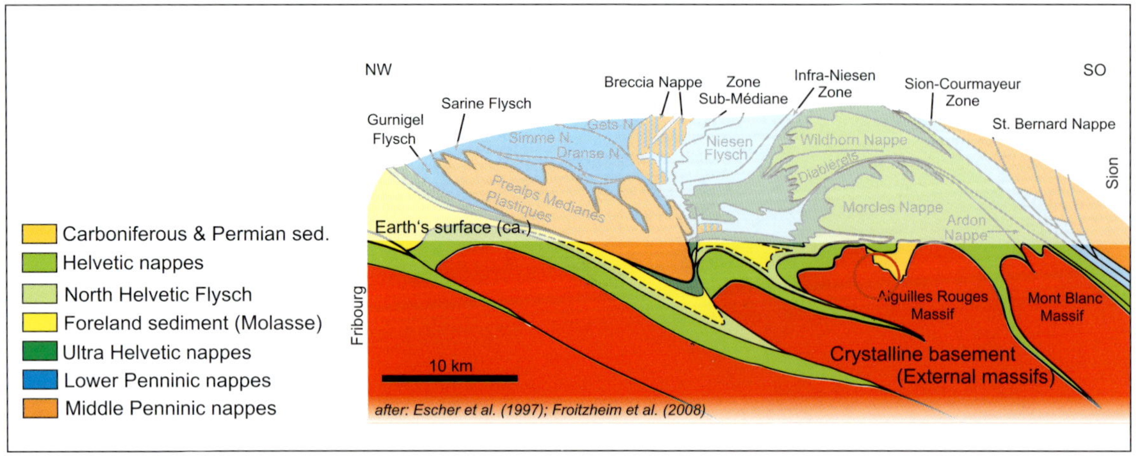

The geological history of the massif is closely related to that of the Mt. Blanc Massif. It also consists of granites and gneisses of the European crust that were formed in Variscan times about 300 Ma ago. However, in the Aiguilles Rouges Massif there is much less granite than in the Mt. Blanc Massif, and most of it is made up of high grade metamorphic gneisses and amphibolites. During the Alpine mountain building of the last 50 Ma, the Aiguilles Rouges Massif was not buried quite as deep as the Mt. Blanc Massif. The massif is surrounded on all sides by Mesozoic sediments (limestone, marl and clays) of the Helvetic Morcles Nappe. On the northern side of the dome-shaped massif the Helvetic rocks lie flat. In the valley of Chamonix they form a steep narrow zone that separates the Aiguilles Rouges Massif from the Mt. Blanc Massif. Helvetic rocks only lie partly in their original sedimentary contact on the Variscan gneisses. In part they are also sheared off along the *Helvetic Basal Decollement*s. During doming and uplift of the massif, about 10 km were eroded off the Aiguilles Rouges, but 15 – 20 km were eroded off Mt. Blanc. 10 km is almost exactly the former thickness of the sedimentary cover. Thus, remnants of these sediments may be found on top of the summits of the Aiguilles Rouges Massif, but not on Mt. Blanc.

Above: The Cime de l'Est at the eastern end of the Dents du Midi near Martigny in the Rhône Valley looking west. The sketch shows the position of the frontal fold of the Morcles Nappe (in orange). The fold hinge is eroded all the way to Tête Bostan.

Right: Historic sketch of the Dent du Midi Massif. It may be seen that the Mesozoic limestone and marl layers of the Morcles Nappe lie on top of Eocene flysch sediments (after: E. Gagnebin, 1925).

The Massif of the Dents du Midi towers high over the west side of the Rhône Valley opposite the Dent de Morcles (p. 24). The geology of the two massifs is very similar. Just like at Dent de Morcles, the Dent du Midi Massif also includes the frontal fold of the Helvetic Morcles Nappe. The nappe consists of Mesozoic limestone and marl and was rolled in and pushed northwards over the flysch zone around 20 – 30 Ma. Only much later did the Rhône Valley cut the nappe into two massifs. The axis of this frontal fold lies parallel to the Val d'Illiez on the northern side of the *Dents du Midi* and the *Dents Blanches*. In the summit region of the Cime de L'Est (3 180 m) the core of this fold may be seen. At the west end of the massif lies *Tête Bostan* (2 407 m). There, the fold hinge may be seen very well, because it is made up of massive Urgonian limestone. In the centre of the massif in the region of Tours Saillere and Dents Blanches, only the lower limb of this giant fold is exposed and the rocks get older and older in higher parts of the range.

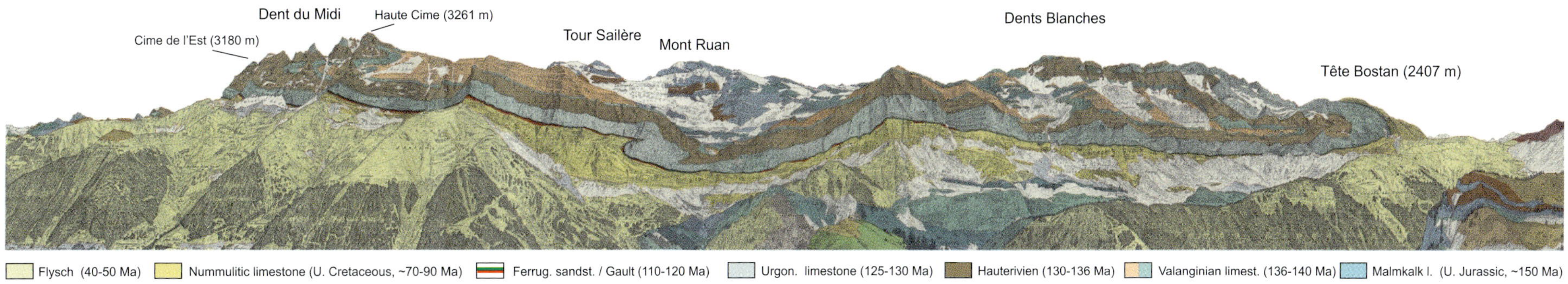

Dent du Midi
Haute Cime (3261 m)
Cime de l'Est (3180 m)
Tour Sailère
Mont Ruan
Dents Blanches
Tête Bostan (2407 m)
Flysch (40-50 Ma)
Nummulitic limestone (U. Cretaceous, ~70-90 Ma)
Ferrug. sandst. / Gault (110-120 Ma)
Urgon. limestone (125-130 Ma)
Hauterivien (130-136 Ma)
Valanginian limest. (136-140 Ma)
Malmkalk l. (U. Jurassic, ~150 Ma)

The front of the Morcles-Nappe is spectacularly exposed on l'Argentine peak east of Gryon near Bex in the lower Rhône Valley. Although l'Argentine (2421 m) is only a minor peak between its huge neighbours Diablerets (3209 m) and Grand Muveran (3051 m), it has an impressive shape and its vertical limestone slabs are well known to climbers. The mountain is made up of shallow marine limestones that were deposited between 100 Ma (Cretaceous) and 35 Ma ago (Oligocene) onto the European continent. Later, tectonic motions scraped the rocks off their crystalline basement and incorporated them into the Helvetic Nappe pile as the so-called Morcles Nappe. L'Argentine forms the frontal fold of the gigantic roll of the Morcles Nappe (p. 230). This fold continues through the Dent de Morcles all the way to the northern slopes of Dents du Midi. It rolled from about 30 Ma onwards from the southeast to the northwest into its present day position. The axis of the fold plunges shallowly to the northeast. Thus, the rock layers become younger and younger near the western end of the l'Argentine. Just below the main summit, the limestone slabs are vertical and near Haute Corde the layered pile becomes more and more upright. The boundary between the Cretaceous limestones and the Nummulitic limestones from the Eocene is marked by a dark coloured horizon of conglomerates and sandstones. This layer mantles the peaks of Haute Point and Haute Corde.

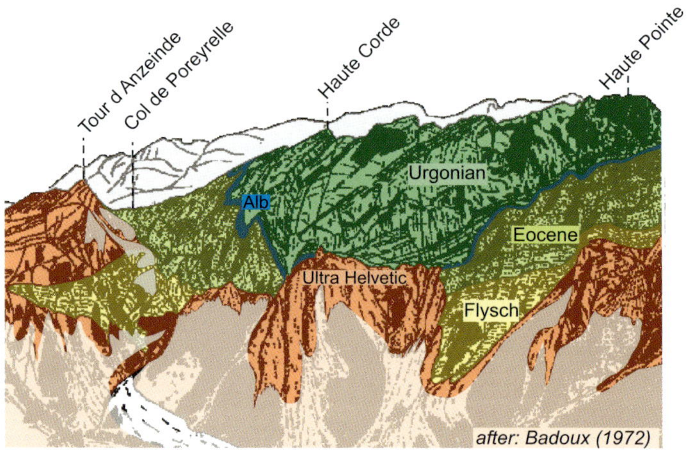

Nummulitic limestone takes its name from a coin-shaped (latin: nummulus = small coin) centimetre-long microfossil, a foraminifera. All rocks of the Morcles Nappe are surrounded by marly limestones and schists of the Ultrahelvetic nappes (Anzeinde Nappe). These were deposited as deep water sediments that were thrust over the top of the Helvetic nappes and then folded together with them.

On the peak Tour d'Anzeinde they lie above the rocks of the l'Argentine; somewhat further out of the valley, they lie below them.

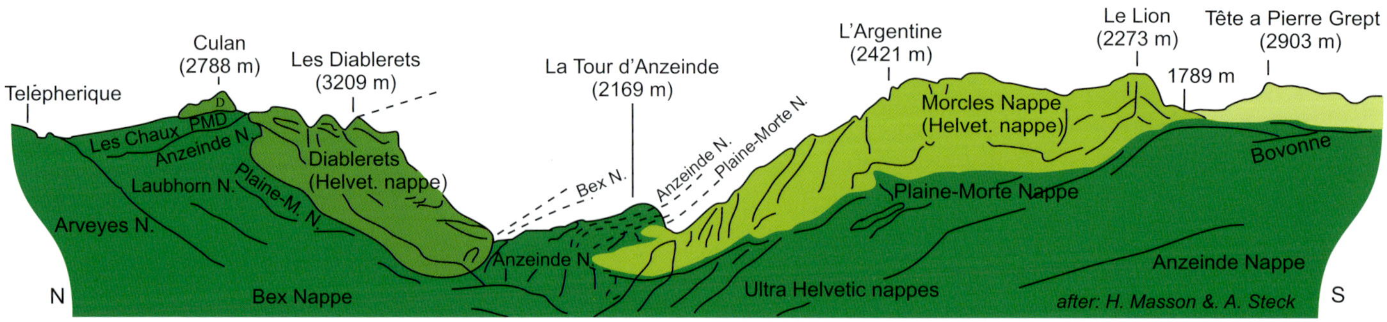

Above: The folded rocks of the Ultrahelvetic nappes in the region of the Tour d'Anzeinde. L'Argentine and the Rhône Valley are in the background.

Right and above right: Two geological sketches of the l'Argentine massif seen from the west.

Page 233: The slabs of L'Argentine from the northeast. The slabs stand vertically because they form the nose of the frontal folds of the Morcles Nappe.

Right: Geological cross section through the Gotthard Base Tunnel. Most of the rocks are part of the Variscan Aare and Gotthard Massifs.

Gotthard Massif and Gotthard Base Tunnel: The Gotthard Massif lies on the south side of the Aare Massif in central Switzerland and consists of Variscan gneisses that are considered to belong to the Sub-Penninic units of the Alps (p. 62). The Gotthard Massif is known for its transit route across the St. Gotthard Pass. This pass (2 108 m) connects the townships of Andermatt (Canton Uri) and Airolo (Ticino) and is the most important pass across the Alps in the Central Alps. Currently, an extremely ambitious tunnelling project is under way in the Gotthard Massif: the Gotthard Base Tunnel will be 57 km long, and upon completion it will be the longest in the world. So far, 7.4 Billion Euro have been spent on the project and 24 million tons of rocks removed from the mountain. The breakthrough between the two sides was achieved in October 2010.

The northern portal of the Gotthard Base tunnel is not far from the south end of the Lake Lucerne. The construction site near Erstfeld and part of the access rails may be seen in the left third of the photo. The inner Reuss Valley behind it lies predominantly in the gneisses of the Aare Massif. However, most rocks in the photo are Mesozoic limestones of the Helvetic nappes, structurally above the Aare massif. The Uri Rotstock (2 928 m) is the peak in the centre of the photo and shows some of the folding in these nappes. The south end of the tunnel is some 57 km further south in Bodio near Biasca in the canton Ticino. The pyramid-shaped mountain at the end o the valley is the Bristen (3 078 m). The photo was taken from about 200 m above the summit of the Fronalpstock (1 921 m).

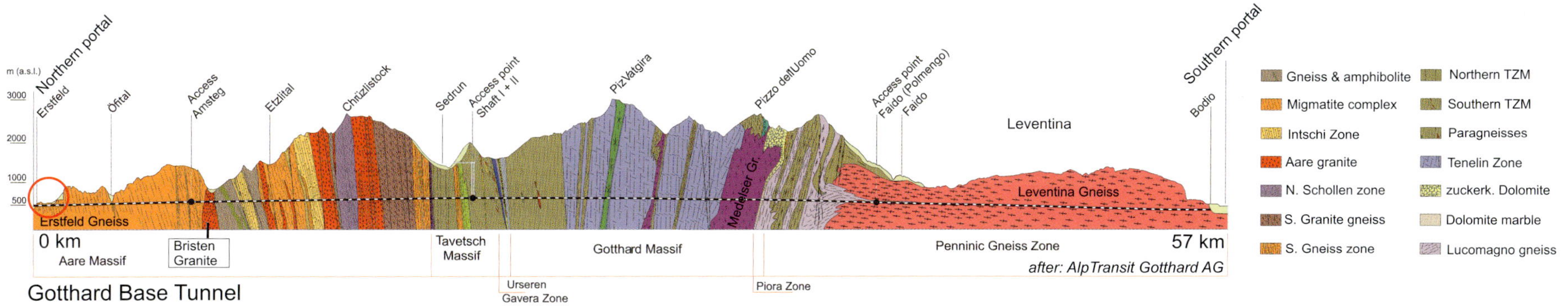

Gotthard Base Tunnel

Crystal Treasures in the Centre of the Alps: Mining for crystals in caves and cracks has the ancient local name of "Strahlen". Strahlen has a century old tradition in the Alps. The granites and gneisses of the 300 Ma old Aare and Gotthard Massifs bear the most famous crystals of the Alps. Perfectly developed crystals of quartz (SiO_2) and fluorite (CaF_2) up to one metre in size have been found here. Bergkristall, smoky quartz, amethyst, rose quartz and citrine are names for differently coloured quartz crystals. Many of these crystals have ended up in museums and university collections around Europe.

Quartz crystals form in open cavities that form due to tectonic motions in the top 10 km of the Earth's crust. At this crustal level, the pressures are small enough so that such cracks can remain open for some time and the temperatures are hot enough so that the crystals can grow. The cracks in the Aare Massif formed about 10 – 20 Ma ago; then the massif started being pushed towards the surface. Hot water that circulated in these cavities dissolved SiO_2 from the surrounding rocks and precipitated it during cooling to form the quartz crystals. In general, the crystals grow from the crack margins towards the open space in the middle, and the best crystals develop if they do not meet from both sides. The last major find in the Alps was made in 2008 on the Planggenstock north of Andermatt.

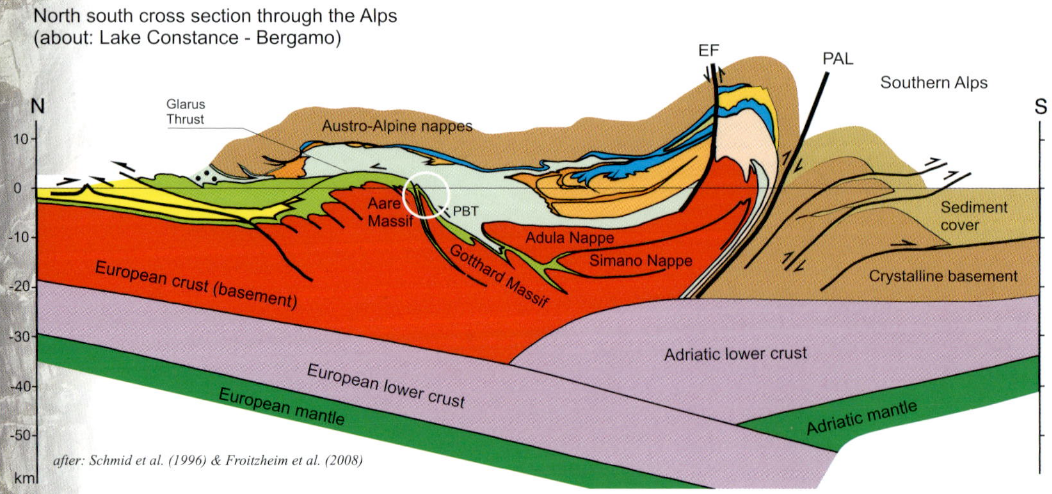

Left: The lonely hut of a "Strahler" (crystal hunter) in a vertical face of the Gotthard Massif.

The geological cross section shows the position of the Gotthard Massif within the Alpine nappe pile (legend on p. 258) (EF = Engadine Fault; PAL = Periadriatic Lineament; ring = Photo this page).

Right: For hundreds of years, the Gotthard Massif has been known for its spectacular finds of quartz crystals. One of the largest finds in the world was made as recently as in 2008 on the Planggenstock.

Right: Geological map of the Préalpes. The small triangles along the Penninic base thrust point in the direction of the tectonically higher slab.

Below: Geological cross section through the Préalpes Romandes in the region of the Gastlosen range.

Page 239: The chain of the Gastlosen mountains seen from the northeast. On the skyline at the very left Mt. Blanc. In the foreground Gummfluh (2458 m). Right of Mt. Blanc the three summits of the Dents du Midi Massif. In the right-hand background Vanil Noir (2389 m), the highest peak of the Fribourg Alps. The two pointy peaks in the Gastlosen chain itself are its two high peaks: Dent de Savigny (2252 m) and Dent de Ruth (2236 m). Behind them Lake Geneva.

The Gastlosen Chain in the Préalpes Médianes: The Préalpes include rocks in the French and Swiss Alpine foreland that once were part of the Penninic Ocean but are now completely detached from the major part of the Penninic Domain of the Alps. Instead, they lie as "free floating islands" (geologists call this "klippe") on top of the rocks of the Helvetic Nappes and the Molasse Zone in the Alpine Foreland. There is a whole series of such klippen starting at the Mythen in the northeast with the southwestern most one being located near Annecy. The two largest klippen are the *Préalpes Romandes* between Lake Geneva and Lake Thun and the *Chablais Préalpes* south of Lake Geneva.

The lowermost unit of the Préalpes is the Eocene flysch of the Niesen. This nappe occurs only in a 5 – 10 km broad band of rocks that starts at the Niesen (2362 m) near Lake Thun and then leads to the southwest past the Giferspitz (2541 m) and the Wistätthorn (2362 m) near Gstaad all the way to the Rhône Valley near Aigle. Northwest of the Niesen-Nappe lies the nappe of the Préalpes Médianes. This nappe is predominantly made up of Mesozoic carbonates. The limestones of this nappe build up most of the peaks in the Fribourg Alps (Gastlosen, Vanil Noir, Gummfluh) and also those of the Haute Savoie region south of Lake Geneva. The uppermost tectonic units are the Breccia and Supérieure Nappes.

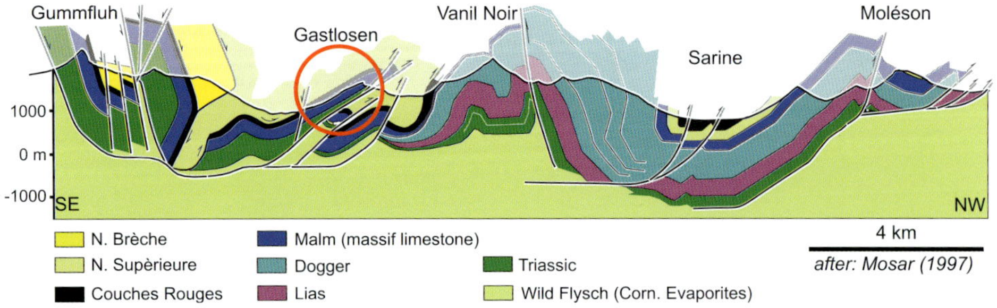

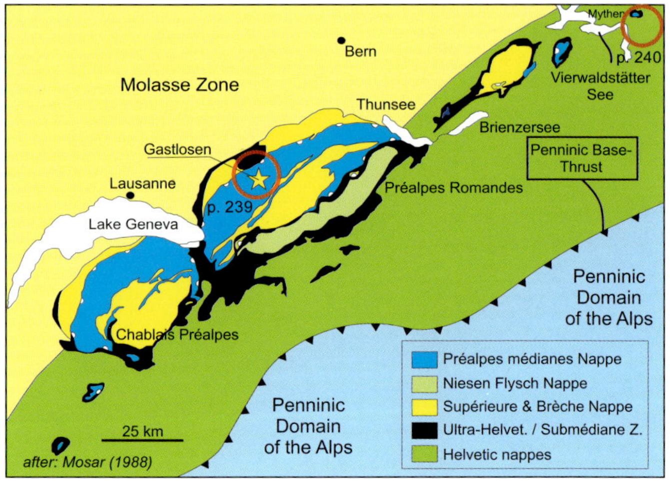

The deformation pattern in the Préalpes is quite different from that in the remainder of the Alps. Brittle fracturing, block rotations, tilting and small shallow displacements are typical instead of large scale folding. Geologists call this type of deformation *thin skinned tectonics.*

The Gastlosen mountains are very typical for the geology of the Préalpes. They form a 15 km long chain in southwest-northeast orientation in the border region between the Swiss Cantons Freiburg, Bern and Waadt east of Lake Geneva. Geologically the Gastlosen are mainly made up of Jurassic Malmkalk limestone. The tectonic motions in the Miocene (10 – 20 Ma) broke the limestone layers into different blocks, and the southeast side was pushed hard against the northwest side, steepening the layers up whilst leaving the northwest side shallow. As a consequence, the steep layers sit between flat lying flysch layers of the same age and the same depositional environment. Within the Gastlosen, the oldest layers are on the northwest side: a characteristic marl-band with mussel fossils that is 170 Ma old.

The Mythen form a free standing massif east of the Swiss city of Schwyz, near the Lake Lucerne. The massif has two summits: the *Kleine Mythen* (1 811 m) and the *Große Mythen* (1 898 m). The main part of the massif is made up of Jurassic reef-limestone (Malm), a rock type that is called "Gebirgskalk" in many other parts of northern Switzerland. The eye catching red summit region is made up of Upper Cretaceous Schieferkalk.

During the Cretaceous the Mythen formed an island in the middle of the Penninic Ocean as part of the Briançonnais Fragment. The limestones that make up the Mythen were deposited at the margins of this island. At around 20 – 30 Ma the Mythen were transported to their present position from a long way south. Today they are a free-floating slab (geologists say: "klippe") that lies atop of much younger rocks, from which they are separated by a shallow thrust zone. Below this thrust zone there are Nummulitic limestones and flysch rocks that were poured into the deep sea trench of the subducting Penninic Ocean about 40 – 50 Ma ago.

The Mythen are the easternmost, but not the only Middle Penninic klippe in the Alpine foreland. For example, the Buochserhorn (1 806 m) and the Stanserhorn (1 898 m) south of the Lake Lucerne, as well as the mountains between Sarnen and Thuner lakes are also rocks derived from the Penninic Ocean and now located in the Alpine foreland. The two largest regions of Penninic klippen lie on both sides of the eastern end of Lake Geneva: the *Chablais Préalpes* in the west and the *Préalpes Romandes* in the Fribourg Alps (p. 238). Each of these regions is more than 50 km long and 40 km wide and the northern end of these regions is being thrust over the top of the Helvetic Nappes and the Molasse Zone.

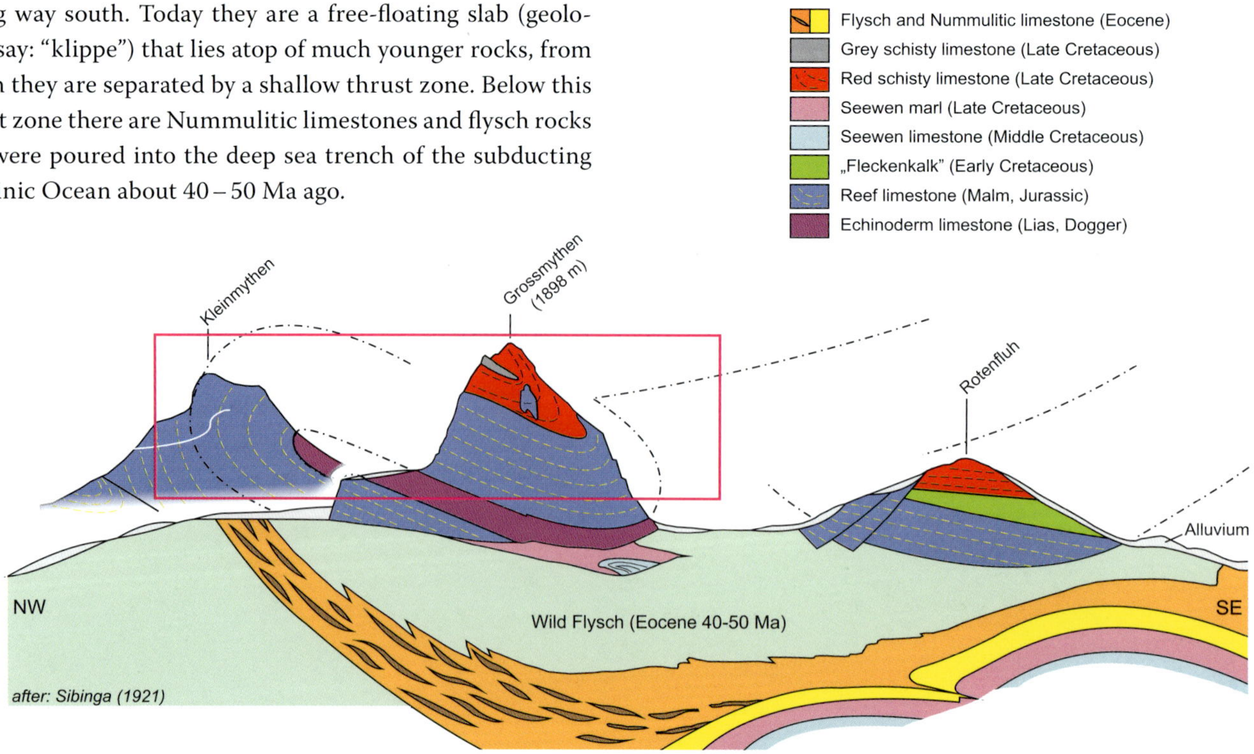

Right: Geological cross section through the Mythen in Switzerland. The red rectangle shows the area of the photo at right. On p. 252 is a more regional geological profile.

Page 241: The Mythen mountains near the city of Schwyz are a tectonically detached slab of rocks (klippe) in the Alpine foreland belonging to the Middle Penninic units. They consist of Mesozoic carbonates that lie on top of much younger flysch rocks.

Above: The Napf region looking east. The summit of the Napf (1 408 m) itself lies in the clouds at the very left. The Swiss city of Lucerne can be seen on the skyline at right.

The Napf region in the Swiss cantons of Bern and Lucerne lies between the famous cheese-maker's valleys of Emmental and Entlebuch. The Napf (1 408 m) is the highest summit of this rolling landscape. The landscape in this region resembles the eastern end of the Alps (for example the Kor- and Saualpe regions p. 104), or the Ötscherland region (p. 138) and does so for a similar reason: these regions are the only regions in the entire Alps that were never glaciated during the glaciation periods (p. 114). During the ice ages, the Napf region sat like an island between the enormous glacial lobes of the Rhône- and Rhine glaciers. For this reason U-shaped valleys and glacial cirques are absent. Curiously enough, the steepest parts of the Napf region are not near the summits, but at the lowest points of the landscape: vertical rock faces are only found near the creeks in the deeply incised valleys. There, creeks flow in narrow gorges directly on bedrock. In contrast, the tops of the

rounded peaks are covered by a thick soil. Farms are located at the high points instead of the low points of the landscape. This observation indicates that the uplift and incision of the region is very recent, probably within the last 5 Ma. Rocks are flat-lying conglomerates (Nagelfluh) and marls that are part of the Molasse Zone of the Alps. The rocks were deposited between 10 and 20 million years ago, partly into fresh water and partly into sea water.

Left: Typical Nagelfluh conglomerate of the fresh-water molasse sediments of the Napf region.

The Eiger: Although the Eiger in the Bernese Oberland is undoubtedly one of the most famous mountains of the Alps it is "only" 3 970 metres high and thus not part of the list of the 82 peaks of the range that exceed 4 000 m in elevation. The cause for its fame is partly because it is so spectacularly located high above the township of Grindelwald (together with its neighbours the Mönch, 4 107 m, and Jungfrau 4 158 m), and in part because of its dramatic climbing history. The Eiger was first climbed in 1858 via its west flank. The impressive 1 650 m high north face was hailed in the thirties of the last century as "the last great climbing problem of the Alps". In fact, the question of who would climb it first became a political issue. A series of teams of Austrian and German origin tried to achive this feat. In 1934 there was a failed attempt, in 1935 two climbers died in a snow storm, and in 1936 the famous drama occurred about which several motion pictures have been made. The Austrian team of Rainer and Angerer and the German team of Hinterstoisser and Kurz joined in the effort to scale the face. After a series of small mistakes and misjudgements all four climbers died on the face. The face was eventually climbed in 1938 by Heckmair, Vörg, Harrer und Kasparek in three days. In 2008 the face was climbed by the Swiss climber Ueli Steck in 2 hours 47 minutes, but the Eiger north face remains reputed for its bad rock and high rock fall danger.

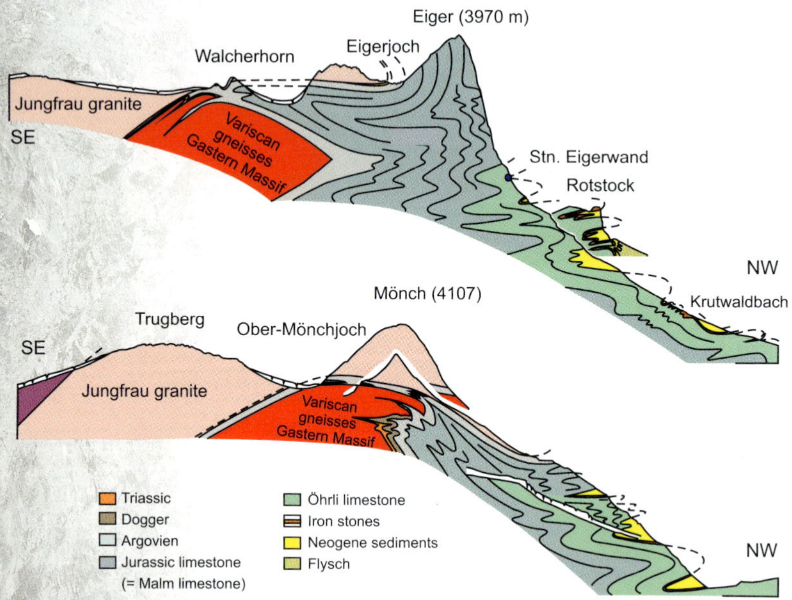

There is a good geological reason for the dramatic accidents on this "killer face". Eiger is made up of the Mesozoic (Helvetic) rocks Quinten limestone (= *Malmkalk*) and Öhrli limestone. The rocks are steeply tilted to the north and lie like a gigantic slab of rock on the northwest side of the Aare Massif. Due to this tilting many of the rock layers are oriented parallel to the face. Because of this surface-parallel orientation, rocks easily break out, and hand and footholds are sloped outwards (in contrast, the limestone layers in the Dachstein south face or in the Dolomites are layered horizontally). At the base of the Eiger north face the Helvetic rocks are abruply bound by marls and sandstones that are part of the Paleogene flysch sediments.

In 2006 one of the largest landslides of modern times occurred on the Eiger. About 500 000 cubic metres of rock fell down from the east ridge onto the lower Grindelwald glacier. A glacial lake was dammed up that threatened the township of Grindelwald for some time.

Left: The Eiger seen from the side in photograph and geological cross section (after: Collet & Parejas, 1931). The limestones that make up the mountain lie like a gigantic slab in front of the Variscan-age gneisses of the Aare Massif just to the south.

Page 245: Eiger seen from the northeast. The location of the 2006 landslide is in the foreground at left. The mountain behind Eiger is the Mönch. On the skyline in the right hand background, Mt. Blanc.

Page 247: The peaks of the Bernese Oberland high above Grindelwald. In the foreground (not against the sky) the Wellhorn (3 191 m). Behind it Wetterhorn (3 692 m) at right and Mittelhorn at left. Behind those, Eiger (3 970 m) and on the skyline the peaks on the west side of the Lauterbrunnental (Gspaltenhorn, 3 496 m). The huge front of these mountains consists largely of Mesozoic limestones of the Helvetic nappes of Switzerland. Due to the uprise of the Aare Massif (in the photo at the very left) these nappes were tilted steeply to the north, folded and finally pushed over the top of the Eocene Grindelwald Basin.

Below: Historical geological sketch of the Wetterhorn.

The Wetterhorn is (besides the Eiger) the second mighty mountain that towers over the town of Grindelwald in the Bernese Oberland. The geology of the two massifs is very similar (p. 244). The huge north faces are made up of Mesozoic limestones that lie on the north side of the Aare Massif (p. 54). Like a slab, several hundreds of metres thick, they lie in front of the crystalline gneisses that build up the core of the Aare Massif. The Schreckhorn and Mittelhorn peaks are slightly set back to the south from the Wetterhorn itself and consist of these gneisses. The summit of Wetterhorn is made of the same rocks as the north faces: limestone. At the base of the north faces, the limestones change abruptly to the flysch and Molasse sediments of the Grindelwald Basin. The limestones, clays and marls of the Wetterhorn Massif were deposited in the Jurassic (about 200 Ma) along the southern margin of Mesozoic Europe into a shallow sea: the northern shore of the Penninic Ocean. Malm- and Öhrli limestone make up the bulk of the north faces. It is the very same rock type that also builds up Tödi (p. 254), the Churfirsten range (p. 260) and many other places in the Helvetic region of Switzerland. When the Penninic Ocean was pushed underneath the Briançonnais Fragment in the Cretaceous, calcareous muds were poured into the deep sea trench. These deep sea sediments lie today in the flysch basin of Grindelwald. Once the ocean was completely gone, the limestones began to be thrust over each other and were folded (p. 244). The sketch below shows that the Öhrli limestone occurs twice in the Wetterhorn: part of it is autochthonous (in contact with its original base – the gneisses of the Aare Massif), the other part is parautochthonous (detached from its base). At about 20 Ma the Aare Massif began to uplift and the limestones were pushed over the top of the Grindelwald Basin and tilted up steeply. Erosion in the last 10 Ma removed about a 1 – 2 km thick section off the soft flysch rocks of the Grindelwald Basin. The Kleine und Große Scheidegg were exposed. The exposure of the huge north faces also occurred in this context and is therefore a matter of only the last few million years, long after the tectonic folding inside the faces.

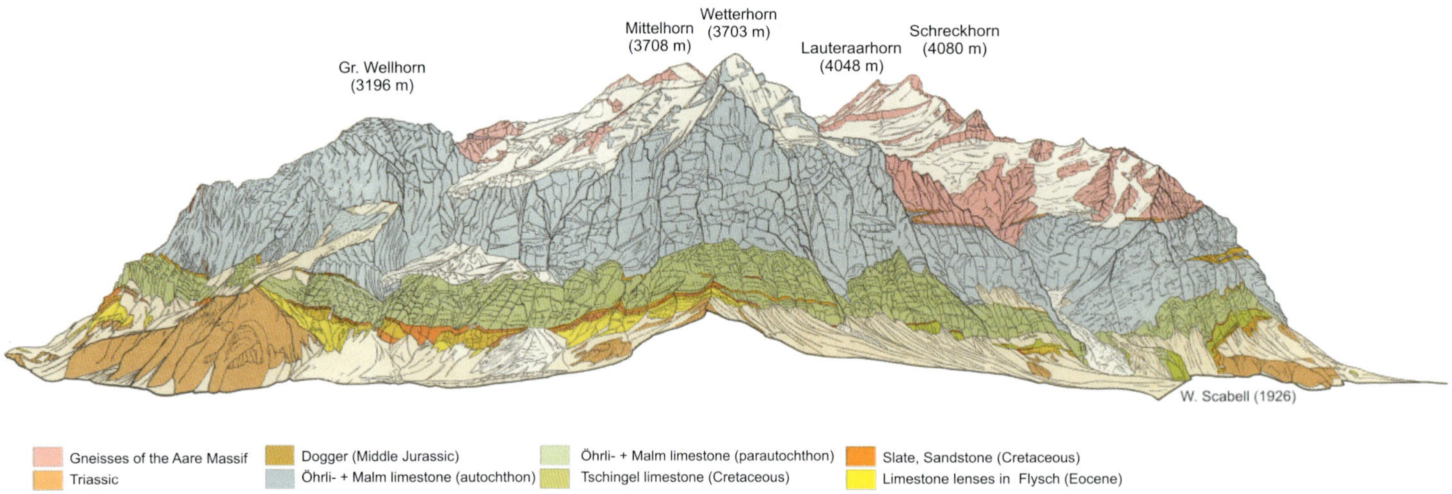

Right: Lago Tremorgio is one of the many small lakes in the Leventina region of the Swiss canton Ticino. Passo Campolungo is seen near the right hand photo margin.

Facing page: Passo Campolungo seen from the east with a view of the spectacularly folded Mesozoic marbles. The large glaciated peak in the middle ground is Basodino (3 272 m), the second highest peak of the Ticino region. On the skyline to its left is Mt. Blanc. The high peaks near the right hand margin of the photo are the mountains of the Bernese Oberland.

The Campolungo region lies around the Campolungo pass just west of the upper Valle Leventina in the Swiss canton of Ticino. Geologically, the region is famous for its spectacular folding of dolomite marbles. The tightly folded marble layers on Campolungo pass itself form one of the best examples for this.

Tectonically, the region is part of the Penninic Zone of the Alps, that is, the rocks were once part of the Penninic ocean floor in the Mesozoic. More specifically, the region is known to geologists as part of the *Maggia* and *Campo Tencia nappes,* which are considered to be part of the Lower Penninic and the Sub-Penninic nappes, respectively (see page 62). The Campolungo pass is located near the northern front of these nappes in a similar structural position as the front of the *Adula nappe* (p. 68). Similar to there, the crystalline rocks of the basement were thrust towards the north over the top of the Bündnerschiefer. The Mesozoic marbles are wedged in between and were folded in the process.

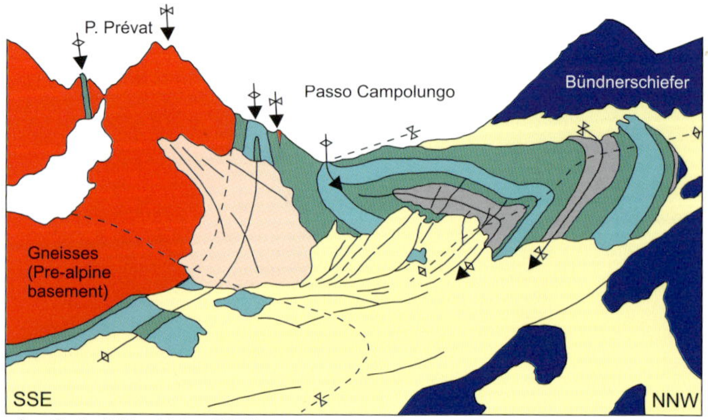

after: Grujic & Mancktelow (1996)

Right bottom: Geological sketch of the Passo Campolungo as seen on the photo on the facing page. Red = Pre-alpine basement of the Lower Penninic and Sub-Penninic Maggia and Campo Tencia nappes; blue = Bündnerschiefer; light blue and greenish colours = different layers of Mesozoic dolomitic marbles.

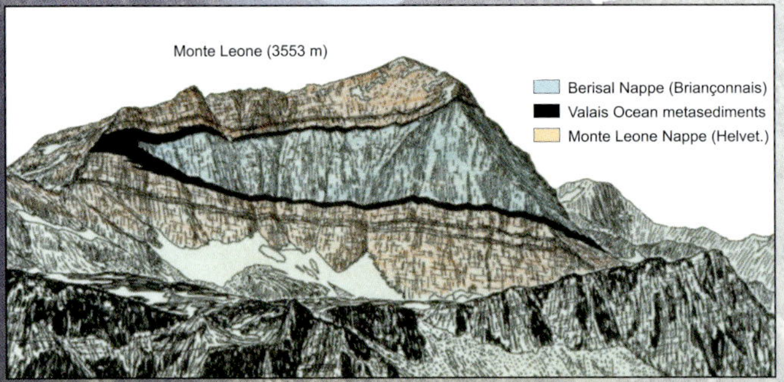

Left: *The west slope of the Seehorn (foreground) and Cima Pioltone (2610 m) (background) are the slip surface of the Simplon detachment near not far from the Simplon Pass in Switzerland. Along this surface, the rocks west of the Pass (in the photo at right) were pulled many kilometres down and to the west (to the right).*

Above: Geological sketch of the photo on the facing page (after: G. Milnes).

Right: Geological profile through the Lepontine Dome. The red square shows the region of the photo at right. The Rhône Simplon detachment dips to the west and therefore appears only as a line on this north-south cross section.

Page 251: Monte Leone above Lago d'Avino seen from the east. The core of the huge fold is made up of oceanic rocks of the former North Penninic ocean floor and of Gneisses of the Briançonnais Fragment.

Monte Leone (3553 m) is the highest peak of the Lepontine Alps in Switzerland. The mountain lies north of the Gondo gorge and east of the Simplon Pass on the border between Italy and Switzerland. Geologically, Monte Leone lies at the west end of the Lepontine Dome, and its impressive west face above Lago d'Avino allows a view of the nappe structure inside the dome. Inside the dome several tectonic nappes are stacked on top of each other.

All of them are part of the Sub-Penninic units of the Alps. The Monte Leone Nappe is the uppermost of these nappes. In the west face of Monte Leone, this nappe is thrown into a huge flat-lying fold. The fold is called a "back-fold" because its hinge closes to the south, although the principal transport of the Sub-Penninic nappes was towards the north. In the core of this back-fold, parts of the North Penninic (Valais-) Ocean and continental rocks of the Briançonnais Fragment are exposed.

The exposure of the complicated folds inside the Lepontine Dome in the Monte Leone West face was made possible through the motion along the Simplon detachment (Rhône-Simplon-Line). The huge detachment surface lies west of Monte Leone. Along this structure, the rocks west of the Simplon Pass were pulled many kilometres off the Lepontine Dome.

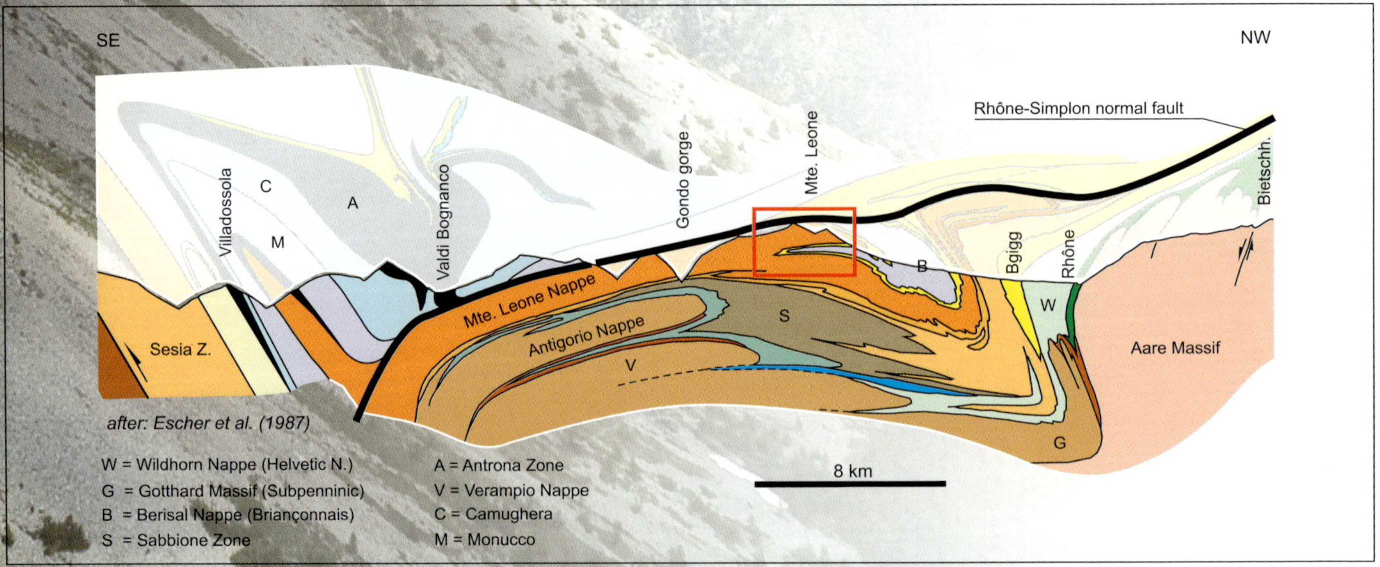

Below: Geological cross section through the mountains east of the Lake Lucerne from the Mythen to the Windgälle. Large ring = photo this page.

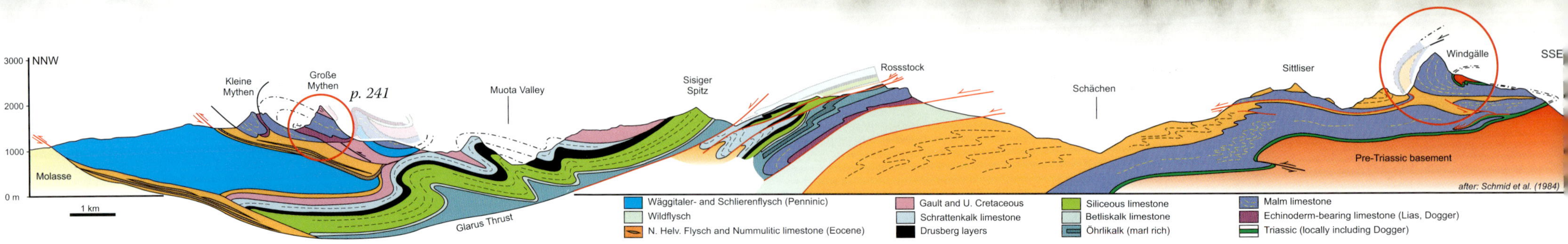

The Windgällen Massif lies east of Lake Lucerne in the Swiss Canton of Uri. Near the summit of this massif, Mesozoic limestones are spectacularly folded together with gneisses from Aare Massif. Up to about 2000 m above sea level, the massif is made up of these 300 Ma old gneisses and granites. Above those follow limestones and marls that are part of the Mesozoic sedimentary cover onto the European crust. However, the summit region of the Chli Windgälle consists of gneisses again. It is the core of a fold with a horizontal axis. The limestones that mantle this fold core may be seen on the Gross Windgälle. The folding occurred 20 – 30 Ma. In space the fold at the Windgälle marks the transition from the autochthon deposited limestones on the Aare Massif (south of Windgälle) to the allochthon limestones of the Helvetic Nappes (further north).

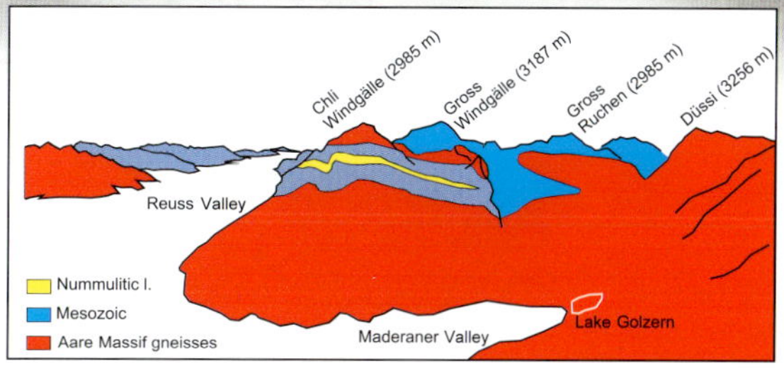

Geological sketch of the photo above.

Valanginian limestone (Early Cretaceous)
„Gebirgskalk" (= Malm = Quintner limestone)
Fe-Oolith, Sandstone (Jurassic, Dogger)
Röti dolomite (Triassic)
Phyllite
Granite
Schist
Aare Massif (Carboniferous / Permian and older)

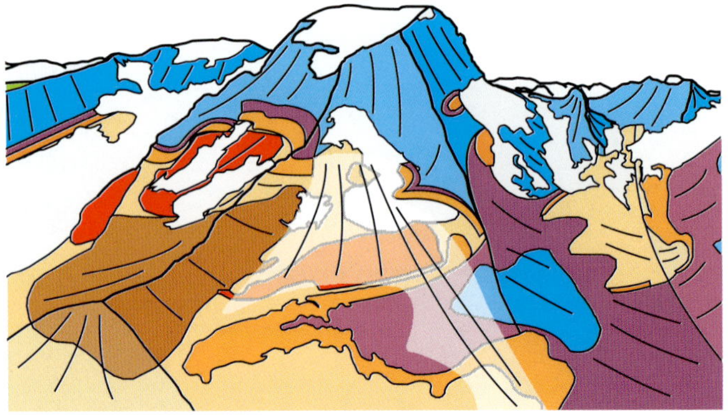

Tödi (3 614 m) is the highest peak of the Glarus Alps in Switzerland. The impressive massif has three main summits: Piz Russein (3 614 m), Glarner Tödi (3 577 m) and Sandgipfel (3 388 m). To the south and north, the valleys of the Rhine and the Linth have incised roughly 2 500 vertical metres deep into the massif in the last 5 – 10 Ma thereby exposing a cross section through a major part of the crust of the Central Alps. The cross section shows the build up of the European crust in the Mesozoic: the lower part of the mountain is made up of crystalline rocks of the Aare Massif that reaches west as far as the Lötschen Valley (p. 56). The rocks are coarse-grained granites and gneisses, but also schists that were deposited as river sediments onto the gneisses in the Carboniferous and Permian. Later these sediments were deformed, folded and metamorphosed into schists. Above these schists, the Mesozoic marine evolution starts that is so familiar from many regions in the Alps, from the Dolomites to the Northern Calcareous Alps: the continent sank and was successively flooded by a shallow sea. The Rötidolomite is a characteristically coloured rock that marks the beginning of this evolution about 200 Ma ago. Above this dolomite various carbonates from the Triassic, Jurassic and Cretaceous follow. Some of these rock types are also known from other parts of northern Switzerland, for example, the massive *Gebirgskalk* or the Lower Cretaceous Valanginian limestone. In the Tödi region the contact between the Rötidolomite and the rocks of the Aare Massif is folded, but the original sedimentary relationships are preserved. The Mesozoic sediments from the upper parts of the mountain are therefore not part of the Helvetic Nappes, but are called the "autochthonous Mesozoic cover of Europe" (p. 58).

Above: Geological sketch of the photo at right. (geology after Weber 1924). The legend is also for the geological cross section on page 257.

Right: Selbsanft (3 029 m) and Bifertenstock (3 421 m) in the Tödi region seen from the west. The orange band of the Rötidolomite separates the gneisses of the Aare Massif (below) from the Mesozoic carbonates (above).

Page 255: Tödi from the northeast from high above the Mutt lakes. The striking orange band is the Rötidolomite.

Page 256/257: The Mutt lakes and the Hausstock (in the clouds) in the Tödi region of the Glarus mountains in Switzerland looking northeast. At left lies the Linth Valley. Behind it in clouds the Glärnisch (2 915 m). The geological profile shows a cross section through the Tödi. The transition from Mesozoic carbonates to the granite gneisses of the Aare Massif is about at the lowest point where the rocks are still sun-lit. The colour scheme of the profile corresponds to the sketch on this page.

NW
Bocktschingel
Creta.
Zementstein beds
Claridenfirn
Spitzalpelistock
Valanginian limestone
Öhrli limestone
Obersand
Schilt beds
Malm breccia
Sandgipfel
Coral limest.
Quinten limestone
Grünhorn series
Tödi granite
Bifertengrätli series
Basal conglomerate
Quartz porphyry
basement
Glarnertödi (NE sum.)
Bifertenfirn
Piz Urlaun
SE
3500
3000
2500
2000

Mürtschen (2441 m) is a mountain on the south side of the Lake Walen in Eastern Switzerland. It may be seen from a long distance and it forms the northern geographic boundary of the Glarus Alps. The north-south oriented summit ridge includes the high points *Ruchen* (2441 m), *Fulen* (2410 m) and *Stock* (2390 m). Geologically the Mürtschen is part of the Helvetic Nappes of Eastern Switzerland and is made up of Mesozoic limestone and marl. During the transport of these nappes towards the north about 20 – 30 Ma ago the limestones and marls were thrown into folds. Due to the alternate layering of light coloured limestone and darker, softer marls, these folds are spectacularly weathered out due to the preferential erosion. The southern part of the mountain is made up of Jurassic rocks (245 – 260 Ma): Quinten limestone in the south (also called *Gebirgskalk* or *Malmkalk*), and Öhrli limestone in the north. The word *Mürtschen* (from "morsch" = rotten) and *Fulen* (foul) are indications for the loose rock. Tectonically the Mürtschen is part of a nappe of the same name. This Mürtschen Nappe is part of the Lower Glarus Nappe Complex that also includes the Glarus Nappe, the Permian volcanics of the Alpine Verrucano and the Axen Nappe. The Lower Glarus Nappe Complex makes up the entire region between Weisstannen Valley, Sernf Valley, Rhine Valley and Lake Walen. To the bottom it is bound by the Glarus Thrust and to the top it is bound by the overthrust Säntis Nappe (p. 260) .

Page 259 and background photo at left: The Mürtschen from the south. Behind it the Churfirsten and the Säntis on the skyline. The highest peak of the Mürtschen is the Ruchen (2441 m), at right.

Below: Schematic geological cross section through Eastern Switzerland. The red ring shows the position of the photo. All rocks above the ring are already eroded. The legend is also for the profiles on p. 32, 62, 73 and 236.

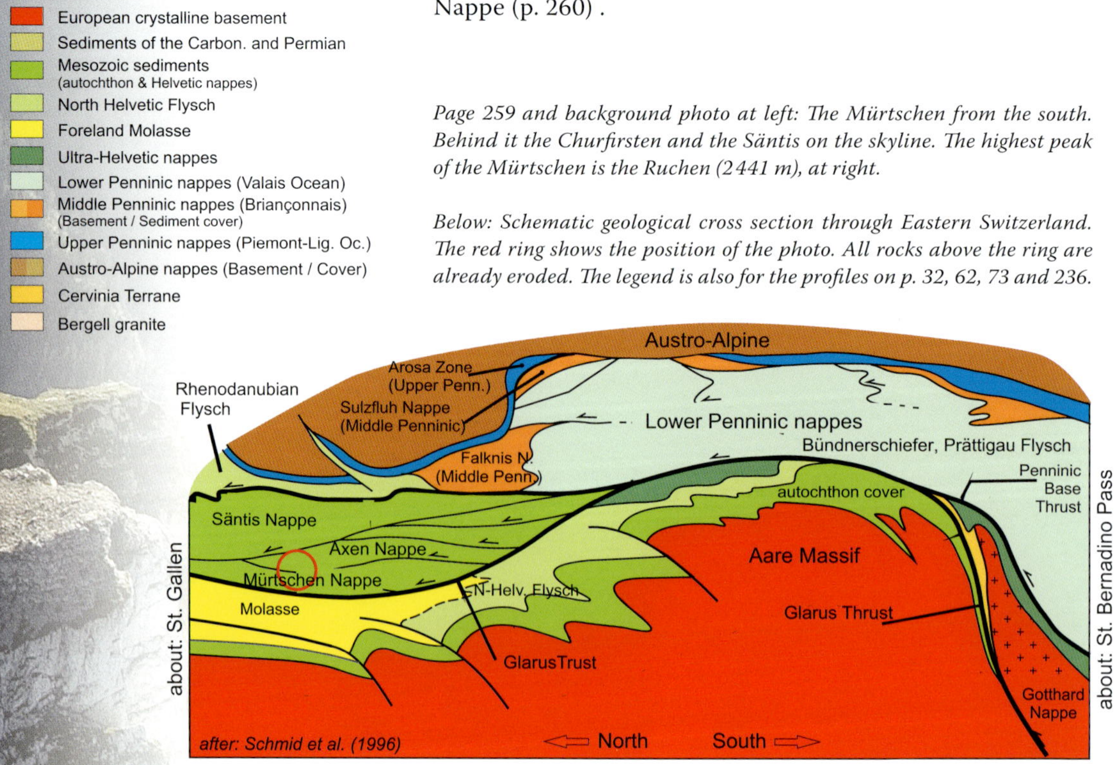

Above: The Churfirsten chain and the Lake Walen. The thrust zone between the Säntis Nappe and the Lower Glarus Nappe Complex lies along the ramp near the settlement of Walenstadtberg high above the eastern end of the lake. The thrust motion has doubled up the massive Jurassic limestone banks.

Right: Geological sketch of the photo above. Stratigraphy after Heim (1924).

The Churfirsten range provides an impressive example for the geology of the Helvetic Nappes of Switzerland. The range lies directly above the Lake Walen along the main route between Zurich and Austria. The rugged shape of the seven main summits of the chain has given the range its name. The range is made up of a horizontally stratified pile of Mesozoic marls and carbonates and is cut diagonally through its middle by a shallow thrust zone. This thrust zone separates the Helvetic Säntis Nappe (above) from the Lower Glarus Nappe Complex (below). The thrust motion itself was part of the deformation of the Helvetic Nappes in the Miocene. Along the thrust zone itself, Eocene sediments may be found that were tectonically dragged in there. Due to the overthrusting, the massive *Gebirgskalk* limestone (from the Malm time about 150 – 160 Ma) can be found twice in the face above the Lake Walen. Firstly, as part of the Säntis Nappe in the centre of the face and secondly, as part of the Lower Glarus Nappe Complex on the level of the lake. One of the most recent layers of the Säntis Nappe is the hard, light coloured Seewen limestone that was deposited in the Upper Cretaceous. Below it there are somewhat older layers of softer, sandy rocks and Schratten limestone (100 – 115 Ma). The rugged skyline of the summit ridge was created due to the layering of harder rocks above softer rocks. The widely visible fold in the Sichelchamm was formed during the folding of the Helvetic Nappes in the Miocene.

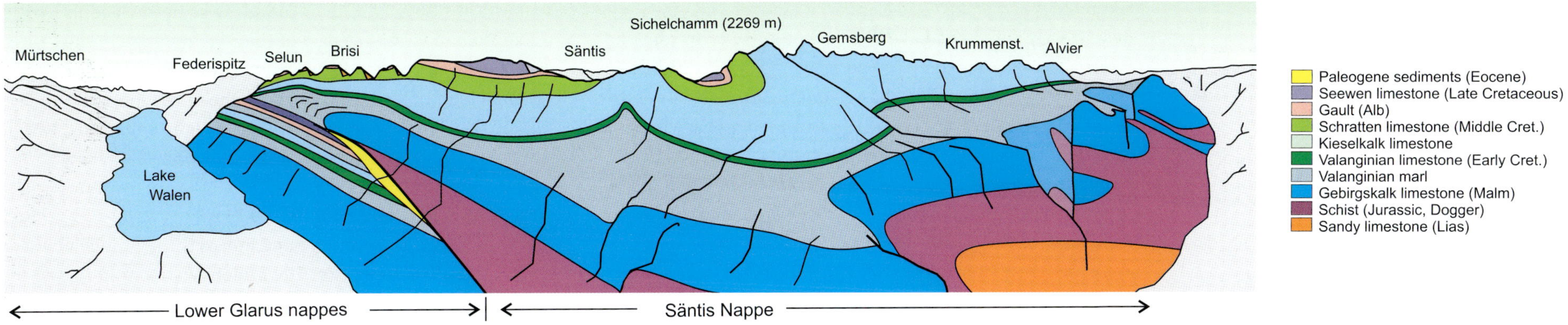

Sichelchamm (2269 m)
Mürtschen
Federispitz
Selun
Brisi
Säntis
Gemsberg
Krummenst.
Alvier
Lake
Walen
Lower Glarus nappes
Säntis Nappe
Paleogene sediments (Eocene)
Seewen limestone (Late Cretaceous)
Gault (Alb)
Schratten limestone (Middle Cret.)
Kieselkalk limestone
Valanginian limestone (Early Cret.)
Valanginian marl
Gebirgskalk limestone (Malm)
Schist (Jurassic, Dogger)
Sandy limestone (Lias)

4.4. Western Alps

The Pelvoux Massif of the Western Alps looking east.

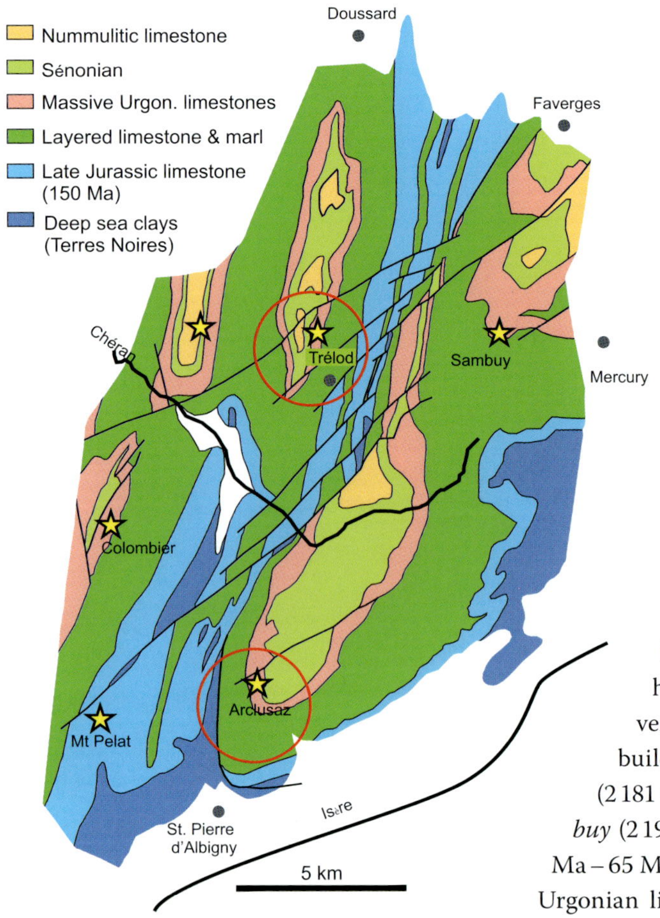

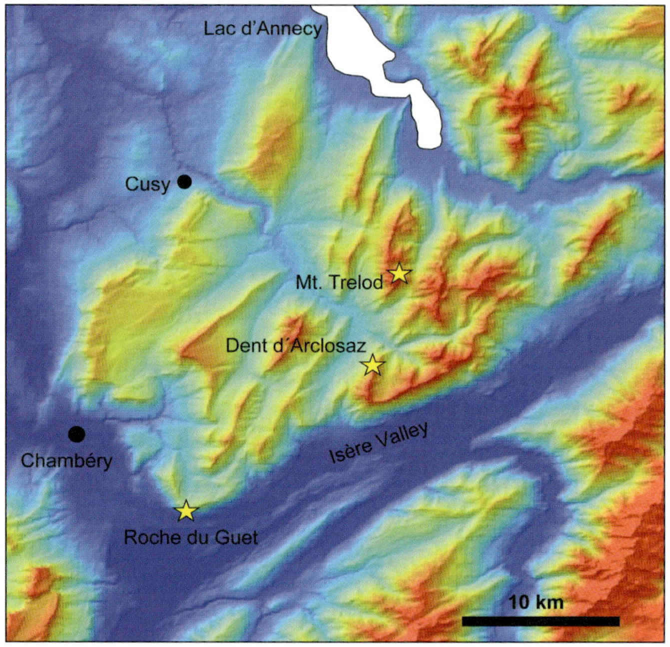

The Bauges Massif lies on the northwest side of the valley of the river Isère between Annecy, Chambéry and Albertville in the French Rhône-Alps. The rocks of the Bauges Massif are closely related to those of the Vercors Mountains or other ranges in the Western Alps. They are Jurassic and Cretaceous sediments that were deposited between 200 Ma and 65 Ma along the northern margin of the Tethys Ocean on top of a crystalline basement of the European continental plate. Thick layers of perfectly white Urgonian limestone (about 125 Ma old and typically several hundreds of metres thick) generally constitute the vertical faces of the peaks. Above those, the summit build-ups of *Dent d'Arclusaz* (2 040 m), *Mt. Trélod* (2 181 m), *Mont Julioz* (1 672 m) and of *Pointe de la Sambuy* (2 198 m) are made from even younger sediments (90 Ma – 65 Ma). In part they bear Nummulite fossils. Below the Urgonian limestones there are usually thinly layered limestones and marls from the Jurassic. During the Neogene, the Bauges Massif was thrown into spectacular, kilometre-scale folds, roughly at the same time that the Belledonne Massif was uplifted to the southeast. These folds strike northeast to southwest and are large enough that they may be seen on satellite images. The Bauges Massif can be subdivided geologically into the *Bauges occidentales* in the west (Mt. Revard, 1 537 m, Mt. Semnoz 1 699 m, Mt. Margériaz 1 845 m) and the *Bauges orientales* to the east. The Bauges occidentales were pushed from east to west over the top of the Molasse basin, and the folds are more open. The eastern part (Trélod, Dent d'Arclusaz, Mt. Julioz) bounds the Isère Valley and includes a quick succession of synclinal and anticlinal folds. Interestingly, the anticlines were later preferentially eroded so that the mountains are made up of synclinal forms while the valleys lie where the former anticlines were.

Above: Geological (at left) and topographic (at right) maps of the Bauges Massif. Even on the topographic map the large scale geological fold structure of the range can be recognised.

The small photo at right shows Dent d'Arclusaz in the foreground and Mt. Trélod in the background. The upper part of both mountains belongs to the same synclinal fold.

Page 265: Mont Trélod (2 181 m) from the south. The spectacular fold that makes up the upper part of the mountain is made of Cretaceous Urgonian limestones. The characteristic limestone band in the core of the fold (left below the summit) consists of Nummulitic limestone from the uppermost Cretaceous. In the background La Tournette (2 351 m) in the Bornes Massif is geologically part of the Préalpes.

The cliffs of the Vercors Massif looking south. The summit in the right hand foreground is the massif of the Grande Moucherolle (2284 m) with its summits, "the two sisters" Agathe (2193 m) and Sophie (2162 m). In the background at left Mont Aiguille (2087 m).

The Vercors Massif in southern France is a nice example of the typical geological build up of the French Alps: far reaching carbonate banks from the Jurassic and Cretaceous are thrown generously into large folds or tilted slightly to the west. These structures were created due to the westwards motion of this part of the Alps in the last 20 Ma. The carbonates are closely related to the Mesozoic cover of Europe as it is found in northern Switzerland (p. 58). However, the rocks here are much less deformed. The base of the Vercors Massif is made from Cretaceous shallow water carbonates that contain more and more evidence for increasing water depth towards the southeast. The vertical faces of the range are built up of Urgonian limestones that lie on this base. Towards the east (left photo margin), the rocks of the massif lie on top of older rocks from the Jurassic and Triassic. East of the valley that connects Vizille and La Mure, the entire Mesozoic pile lies on the crystalline basement of *Belledonne Massif.*

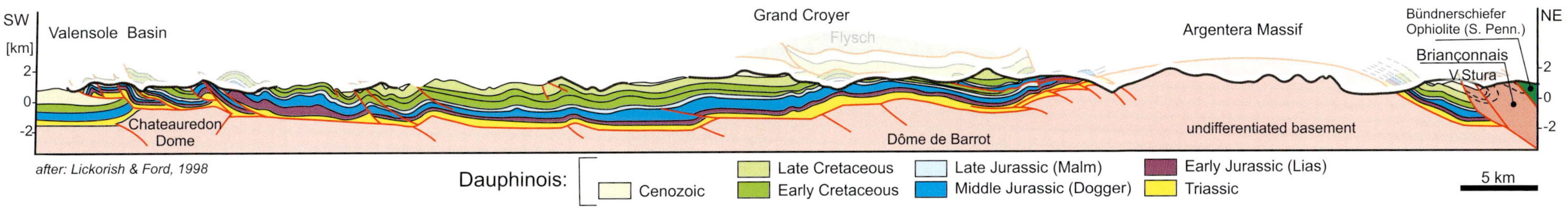

Geological cross section through the Western Alps from the region of Aix-en-Provence across the Alps to Cuneo in the Po plains. Although the profile is south of the Vercors range, it is still characteristic for its structure and in fact for the structure of the entire Western Alps. Kilometre-scale open folds of Jurassic and Cretaceous carbonates and marls dominate the landscape.

Mont Aiguille (2087 m) is located near the highest peak of the Vercors range (Grand Veymont, 2341 m). In comparison to its neighbours it is not very high, but because of its curious climbing history it is still one of the most famous mountains of the Alps. The mountain was long considered unclimbable until King Charles VIII ordered its ascent in 1492. The group of climbers was led by the king's chamberlain Antoine de Ville. This ascent is considered to be the birth of Alpinism in the Alps.

The geological build-up of Mont Aiguille is typical for the entire Vercors range. In fact, it can be used to illustrate the geology of much of the French Alps. The mountain sits on a base of Jurassic deep sea sediments (= *Terres Noires*, not exposed at Mont Aiguille). Above those, there is a thick pile of puff pastry-like alternations of marls and limestones. This unmistakable sequence was deposited in the Lower Cretaceous (Hauterivium, 130 – 135 Ma) and provides evidence for deposition in alternating deep water (marls) and shallow water (limestones). Above this sequence the summit build-up of the mountain is made up of massive limestone (Lower Cretaceous: Barremium to Abt, 115 – 130 Ma). This bright white limestone is also called Urgonian limestone and roughly corresponds to the Schratten limestone of the Helvetic Nappes of Switzerland. On Mont Aiguille erosion and incision of the surrounding valley have isolated these Urgonian limestones like an island. Further west, they build up Grand Veymont and continue from there all the way to the Rhône Valley near Valence.

Below: The grassy plateau on the top of Mont Aiguille. In 1957 a plane landed here. Behind it the highest peak of the Vercors range: Grand Veymont (2341 m)..

Above: Tête de Garnesier (2367 m) and Roc de Garnesier (2383 m) in the Dévoluy Massif northwest of Gap show the pastry-like pile of thinly alternating marl and limestone layers from the Cretaceous.

Page 269: Mont Aiguille from the south. The thinly layered base of alternating marls and limestones from the Lower Cretaceous (130 – 135 Ma) is easy to recognise. Above it the massive Urgonian limestones. The background image shows a detail of the same rocks in the cliff of Roc de Garnesier.

Geological map of the Western Alps

Rocks of the South Penninic & Austro-Alpine

- "Schistes Lustres" (Bündnerschiefer)
- "Breche Nappe" (External Piem.)
- Flysch (Helminthoid)
- Internal Massifs (Sub-Penninic)

Briançonnais Fragment

- External Zone (Carboniferous sed.)
- Internal Zone (Basement)
- Prealps (Sub Briançonnais)

Dauphinois (European cover)

- Valais Zone
- Ultra Dauphinois, Helvetic
- Dauphinois Zone (Triassic / Jurassic)
- External Massifs
- Sub- Alpine (Cretaceous sediments)

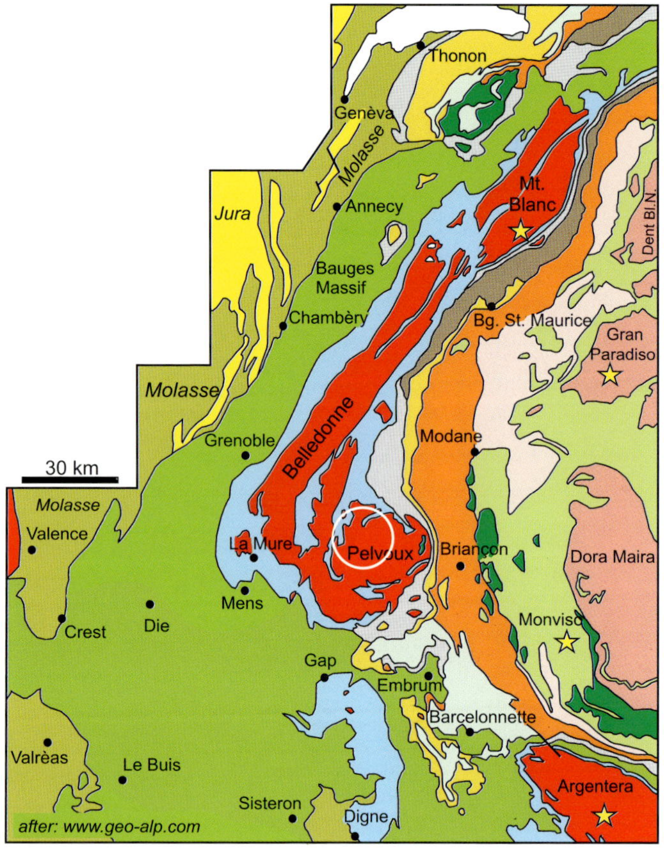

The Écrin Massif in the French Dauphinois region lies between Gap, Grenoble and Briançon. The geological name for the massif is the *Pelvoux Massif.* It is one of the five External Massifs of the Alps in which the crystalline basement of the European plate was domed up by the tectonic processes of the last 20 Ma and pushed towards the surface through the overlying sedimentary cover. These Mesozoic sediments surround the massif on all sides (Dauphinois Zone). In the northwest, the massif is connected to another one of the External Massifs, the *Belledonne Massif.* The Pelvoux Massif is about 40 km in diametre and includes the highest peaks of the Dauphinois Alps: Mont Pelvoux (3 946 m, formerly considered to be the highest peak of the region), Barre des Écrins (4 102 m) and La Meije (3 983 m). Écrin itself is the southernmost and at the same time westernmost peak of the Alps that exceeds 4 000 m in elevation. Geologically the Pelvoux Massif has a large Variscan granite in its centre. In the contact aureole of this granite, the gneisses were partially melted at about 300 Ma. Today, these rocks are migmatites and high grade metamorphic rocks.

Above: Geological map of the French Alps (ring = photos this and facing page).

Right: La Meije (3 983 m) from the south seen over the top of the west ridge of Écrins. The double summit behind La Meije are the Aiguilles d'Arves (3 514 m).

Page 271: La Meije (3 983 m) from the north. The mountain at right is Le Râteau (3 809 m). To the right behind this summit is the French ski resort La Grave. In between, the highest peak of the Pelvoux Massif, Écrin (4 102 m), may be seen in the background. All mountains are made up of granites and gneisses of Variscan age.

Right: Endless mountains that are made up of oceanic lavas of the former Piemont Ocean form the mountains northwest of the Dora Maira Massif in Italy.

Below: Geological map of the French Alps in the region of Monviso and the Dora-Maira Massif. The black line shows the position of the geological cross section on the facing page (rectangle = photo on p. 273).

Page 273: Monviso (3 841 m) from the northeast. On the very right the light coloured layers of the Crête de la Taillante (3 197 m). There, the Bündnerschiefer rocks of the Queyras Complex lie on top of the serpentinites of Monviso. Near the left photo margin, the rocks of Monviso lie atop the Dora-Maira Massif. Rifugio Barbara in the foreground.

Monviso (3 841 m) is one of the most impressive mountains of the Western Alps. It is more than 500 m higher than any other mountain in the region and stands directly above the Po plains southwest of Torino. Geologically, Monviso (and the neighbouring *Queyras Schistes Lustrés)* consist of rocks derived from the Jurassic South Penninic (Piemont-Ligurian) Ocean. The rocks derived from this ocean build a 30 km wide band of rocks that starts at the Monte Rosa Massif, continues beneath the Dent Blanche Nappe, goes around Gran Paradiso and leads south to the Monviso region until it disappears underneath the Po plains near Cuneo. Monviso itself is made up of serpentinites, gabbros and oceanic lavas. During the later subduction of this ocean the rocks were dragged down to more than 70 km depth and then metamorphosed about 50 million years ago to eclogites.

Only thereafter were the rocks squeezed to the surface along large thrust zones.

To the east, the Monviso complex lies on top of the Dora-Maira Massif, one of the three Internal Massifs of the Alps. There, the rocks are of continental origin and are allocated a Sub-Penninic tectonic position. During the subduction of the neighbouring ocean, the Doara-Maira Massif was dragged down to more than 100 km depth and thus contains high pressure minerals that are found nowhere else in the Alps.

To the west, the rocks of the Queyras Schistes Lustres complex lie atop the Monviso Complex. The Queyras Schistes Lustres are made up of Bündnerschiefer and are also part of the Mesozoic Piemont-Ligurain ocean. Its Bündnerschiefer are sediments of the former deep sea trench. During subduction, these sediments were pulled to depth, and parts of the former ocean floor lavas mixed with them and metamorphosed about 55 Ma. Today, the rocks contain lawsonite, glaucophane and other minerals that are characteristic for blueschist facies metamorphism (p. 84).

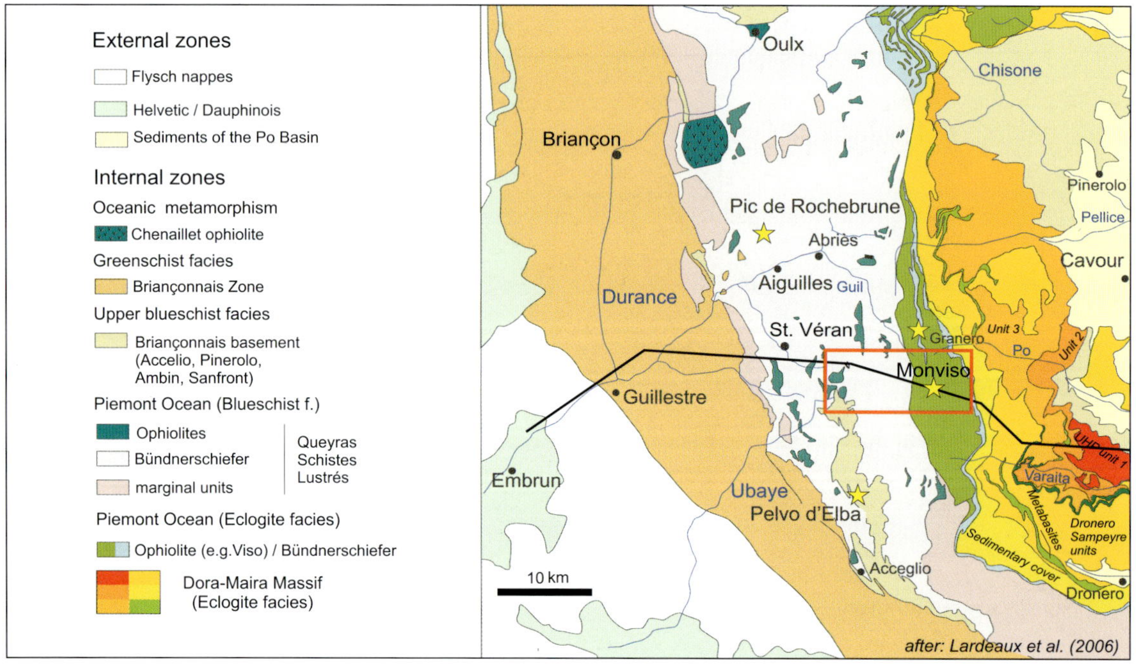

DORA MAIRA
Monviso
BÜNDNERSCHIEFER
(Piemont Schistes Lustrés)
BRIANÇONNAIS
EXOTIC FLYSCH NAPPES
Monte Ricordone
Punta Seras
Le Queyron
Cascavelie
Col de
Bramousse
Sommets
d'Assan
Durance
Pic du Clocher
E
W
2000
1000
Po Plain
Gypsum
Rhyolite
2 km
after: Lardeaux et al. (2006)

Above: Between the townships Guillaumes and Daluis in the French Alpes Maritimes, tectonics and erosion have opened up a window through the limestone sequences of southern France: the Dôme de Barrot. There, Permian red beds are exposed from underneath the Mesozoic carbonates. The river Var has incised a spectacular gorge through these sandstones that is a popular tourist destination.

The Geology of Southern France is characterised by Triassic and Jurassic limestone sequences, similar to the *Vercors Massif* (p. 266) or the *Gorge de Verdon* (p. 278). As for these regions, the Mesozoic sequences form a thick cover on top of the basement of the European plate. In the *Provence region,* these limestone sequences were detached from their base along the *Digne Thrust System* and pushed westwards over the top of the Miocene rocks further west (see cross section on p. 267). The thrust system runs roughly north-south through the city of Digne from where it bears its name. Behind the thrust (i.e. further east), remnants of Eocene foreland basins rest on top of the Mesozoic limestones. Even further east, in the vicinity to the valley of the *river Var,* a large window opens up through the limestones to expose the view onto Permian rocks that lie underneath the Mesozoic sequences: the *Dôme de Barrot.*

In the last few million years of the geological history much of the Provence region experienced a renewed young uplift event. Because this surface uplift is so young (and glaciers were never present in the region), rivers have not found time yet to dissect the region. Thus, many of the mountains feature plateaux in their summit regions (similar to the Koralpe region, see p. 104). The *Valensole* and the *Chambaran* plateaux are typical for this.

Facing page: Jurassic and Eocene limestone sequences in thrown into generous open folds dominate the picture of southern France. The image shows the village Thorame-Haute in the upper Verdon Valley. The highest peak in the background is the Tête de l'Estrop (2 961 m), the highest peak of the Massif des Trois-Évêchés in the French Alpes de Haute Provence.

Bottom right: The small town of Tallard south of Gap in the French Dauphiné region is one of the many medieval townships of southern France that give the region its peaceful atmosphere. Tallard also has an airfield that served as a base for many of the photos shown on these pages.

The Argentera Massif is the southernmost of the five External Massifs of the Alps, with the others being the Mt. Blanc, Belledonne, Pelvoux and Aare Massifs. The Argentera Massif lies southwest of the Italian city of Cuneo in the Alpes Maritimes, in the border region between France and Italy. The massif is about 50 km long and 30 km wide and lies in a northwest – southeast orientation along the principal divide of the Alps. As for the other External Massifs, the Argentera Massif is an updoming of the European crystalline crust and is exposed in a window like fashion from underneath the surrounding Mesozoic sediments. The rocks are ancient granites and gneisses that were last metamorphosed and intruded by granites during the Variscan orogenesis about 300 Ma ago. During the Neogene Alpine processes the massif was hacked up by shear zones and the northern part pushed somewhat over the top of the southern part. The final exhumation of the massif towards the surface began about 25 Ma ago and was finished only 1 – 2 Ma ago. During the glaciation periods of the last 100 000 years, the rugged landscape of glacial cirques and the many lakes were created. The highest peak of the massif, Cima Argentera (3 297 m), is also the highest peak of the Alpes Maritimes.

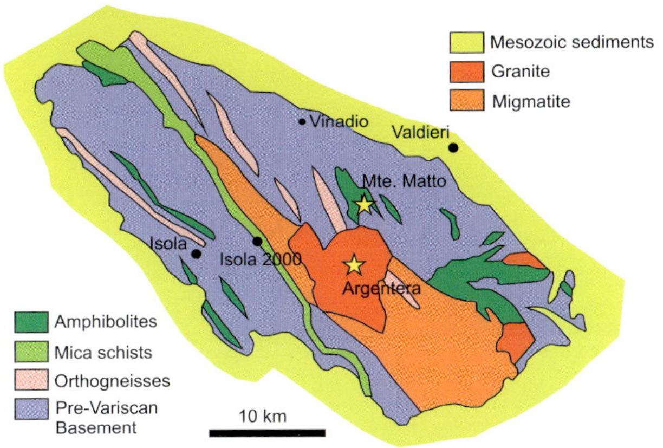

Above: Panoramic view of the central Argentera Massif, the southwestern most of the External Massifs of the Alps. The highest peak of the massif is Cima Argentera, the summit in the centre of the left hand side of the photo high above the settlement of Terme di Valdieri. In the background, the valley leads to the township of Valdieri in the Italian Gesso Valley. The three lakes at right are the Lago delle Portette, Lago del Claus and Lago Vallescura. Behind them, the ski slopes of the famous French ski resort Isola 2000. From there, the massif drains towards the French Tineé- and Vesubié Valleys.

Left: Geological map of the Argentera Massif in the French-Italian border region of the Alpes Maritimes (after: J. von Raumer et al., 1993).

The gorge of Verdon seen from the south. The vertical limestone face is called the "Barre de l'Escalès" and is considered one of the best rock cimbing cliffs in Europe. The peak on the skyline is the Petit Mourre (1 873 m). The township of La Palu lies in front.

The Gorge of Verdon – Grand Canyon of the Alps: In the Provence region, the Verdon river flows through a 700 m deep gorge across the limestone plateaux of southern France. In this distal part of the Alps, the Mesozoic limestone layers that cover the European crust are less deformed than in much of the Alps. In that respect, the Provence region resembles the Vercors mountains (p. 266) or even the Swiss Jura mountains. Often the layers are only somewhat tilted to the west. The westwards inclination of the *Plateau Valensole* on the north side of the Verdon river shows that this tilting continues into recent times, because the plateau is not completely dissected by rivers yet.

The rocks of the Provence region were deposited in the Jurassic and Cretaceous along the northern margin of the Tethys Ocean. During the subsequent Neogene tilting and breaking of the carbonate plateau, the Verdon river found its way through the soft limestones. However, the gorge itself formed only in the last 2 Ma during the glaciation periods. During this time, the Verdon river grew to an enormous river that drained

the glaciers of the Alps to the southwest. Geologists estimate that the river carried up to 3000000 litres of water per second (more than the river Rhine today), incising the gorge at an incredible rate. Today, the *Barre de l'Escalès* in the Gorge de Verdon is one of the most famous rock climbing areas in the world. The face is made up of 150 Ma old massive limestone. Below this there are thin layers of dolomite and limestone that contain brachiopod fossils. The lowermost layers near the river are up to 170 Ma old.

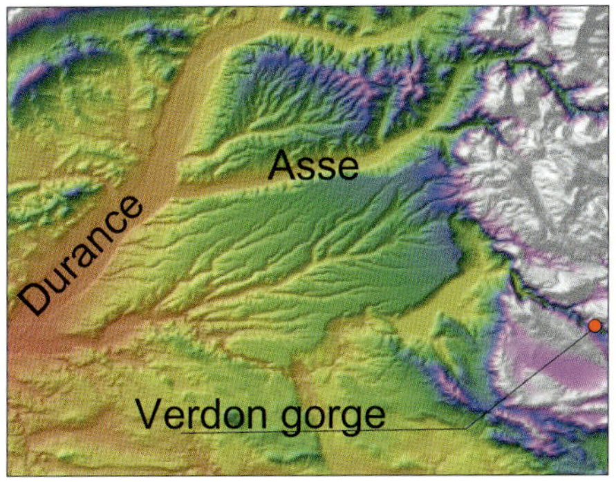

Left: Digital elevation model of the Valensole plateau (the plateau east of Durance and south of the Asse in the image centre) and the gorge of Verdon at the southwest margin of the Alps. It may be seen that the Valensole plateau is tilted slightly to the west and is not completely dissected by rivers.

Glossary

Alpine Mountain Building: The sum of all geological processes of the last 150 Ma in the Alps. The term differentiates the Alpine orogeny from the *Permian* and *Variscan* events. In the Eastern and Southern Alps, the Alpine events are subdivided into *Eo-Alpine* (Cretaceous) and *Young Alpine* (Paleogene) events. In the Central and Western Alps only the latter event occurs. The Paleogene events are thus often referred to as the *Alpine event* in the narrow sense.

Basement: The word usually refers to the crystalline rocks that form the base of a sedimentary sequence. However, more generally, any underlying rock can be called basement with respect to the overlying rocks under discussion.

Breccia: Cemented rock made up of angular fragments. Breccias are typical evidence for tectonic faults.

Conglomerate: Cemented rock made up of rounded pebbles. Conglomerates are typically fossilised river beds.

Continent: A region of continental lithosphere. Normal continental lithosphere consists of about 30 km crust plus some 100 km of the upper mantle that are colder than about 1 200 °C.

Crust: The chemically differentiated outer shell of Earth made from partial melts of the mantle. On the average, the crust is of granitic composition. *Continental crust* is 30 – 40 km thick. *Oceanic crust* only 5 – 7 km.

Detachment: Brittle or ductile shallow-dipping zone along which the rocks above this zone are removed from the rocks below by stretching. Detachments are typical structures forming during continental extension.

Ductile deformation: Permanent, soft deformation of rocks, for example folding. Ductile deformation includes a series of processes. The term serves to contrasts *ductile* from *brittle* and *elastic* deformation processes.

Eclogite: Silica-poor rock of intrusive or volcanic origin that was transformed by high-pressure metamorphism to a rock made of *garnet* and *omphacite*. The minerals zoisite, kyanite, amphibole and plagioclase often occur in minor amounts.

Facies: *Metamorphic* facies are rough ranges of pressure and temperature in which similar groups of minerals grow in rocks of similar bulk composition. *Sedimentary* facies describe related sedimentary environments.

Flysch: A deep sea sediment with signs of active deposition. Flysch sediments are made up of marly, calcareous clays and sand. Such sediments were poured into the kilometre-deep ocean trenches that form in connection with subduction zones.

Gneiss: High grade metamorphic, deformed, coarse-grained rock. *Paragneisses* form from sedimentary rocks, *orthogneisses* form from former intrusive rocks.

Klippe: Part of a tectonically displaced package of rocks that is mostly eroded, except for an island-like raft that lies on top of tectonically foreign rock: the klippe.

Lithosphere: The shell of Earth that is made from rock and drifts as tectonic plates on top of the much softer asthenosphere. Two types of lithosphere are identified: *continental* and *oceanic*. Both consist of the Earth's crust plus several tens of kilometres from the upper mantle which are cold enough to be part of the rocky plate.

Mantle: Part of Earth between the crust (about 30 km depth) and the core (at about 3000 km depth). The mantle consists largely of olivine ($MgSiO_3$). The uppermost 50 – 100 km of the mantle are part of the lithosphere.

Messinian crisis: A 500 000 years long period between 5.4 and 5.9 Ma during which the straits of Gibraltar were tectonically closed. The Mediterranean sea almost dried completely out.

Metamorphism: The change of mineralogy, grain size and grain shape in rocks due to changed pressure and temperature conditions.

Mountain Building: The sum of all geological processes that occur during collision of two continents, including structural development, metamorphism of rocks and – of course – also the mountain building in the narrow sense: uplift of the surface to form a mountain range.

Ocean: Region of oceanic lithosphere. Oceanic lithosphere is usually thinner, stiffer and more dense than continental lithosphere. Therefore the surface of oceanic lithosphere lies lower and is often covered by water (see: *Sea*). When oceanic lithosphere collides with a continent it is generally subducted underneath it.

Ophiolite: Term describing a complete cross section through the oceanic lithosphere that was tectonically squeezed to the surface. A bit more loosely, the term is often used to refer to any splinter of oceanic crust that is found in a mountain belt.

Orogeny: A very general term used to refer to plate tectonic scale geological processes at plate margins. The term *orogeny* is often used synonymously with *Mountain Building Processes*, but orogeny also includes extensional plate margin processes.

Pegmatite: A magmaic rock that forms usually as the first or last melt during melting or crystallisation processes, respectively. Pegmatites typically intrude as dikes. Pegmatites are usually of roughly granitic composition but because of their origin as "left over" melts, they typically contain large amounts of incompatible elements and often contain gemstone deposits.

Pull-Apart Basin: Small sedimentary basin that was pulled open by fault zones and is bound by these on all sides.

Sea: A salt water covered part of Earth. Usually a sea is underlain by oceanic lithosphere, but sometimes a sea forms on a continent, for example the Adriatic (see also: *Ocean*).

Strike-Slip Zone: Steeply inclined narrow zone along which the rocks from both sides are displaced with respect to each other. The San Andreas Fault is the most famous strike-slip zone on Earth.

Subduction: The submergence of one lithospheric plate underneath another. Typically, oceanic plates are subducted underneath continental plates and a deep ocean trench is formed. At depth, high-pressure metamorphism of rocks occurs in subduction zones.

Suture: Tectonic scar on a large scale, for example of a former plate boundary.

Thrust Zone: A brittle or ductile zone that is usually inclined with less than 45° along which the upper plate is moved over the lower plate. Thrust zones are clear evidence for compressional tectonics.

Variscan Orogeny: A tectonic phase of mountain building that affected large parts of Europe some 300 Ma ago. In the Alps, many rocks formed at this time were reworked during the Alpine orogeny.

Window: A region where tectonics and erosion have removed the upper layers of the crust so that lower lying layers are now exposed. However, the upper layers must be preserved in a frame-like fashion on all sides of the window.

References

Overview articles and books

The Geology of Central Europe. Vol. 2: Mesozoic and Cenozoic. Geological Society of London. McCann T. (ed.) (2008). *Therein*: Froitzheim et al., Alpine Tectonics of the Alps and Western Carpathians, Chap. 18, pp. 1141–1232.

Tectonic Map and Overall Architecture of the Alpine Orogen. Schmid S et al. (2004). Eclogae Geologicae Helvetiae, v. 97, pp. 93–117.

Beiträge zur geologischen Karte der Schweiz. *Therein:* Heim (1924, *Churfirsten*); Keller (1910, *Bifertenstock*); Gagnebin (1925, *Dents du Midi*); Sibinga (1921, *Mythen*); Scabell (1926, *Wetterhorn*); Weber (1924, *Toedi*).

Geologie der Schweiz. Heim A (1921). Band II. 1, Tauchnitz, Leipzig. *Therein: Rawil Depression.*

Geology of Switzerland Part A and B. Trümpy et al. (ed.) (1980). Schweizerische Geologische Kommission & Co Publishers, Basel, New York *Therein*: Steck (1979, *Aare Massif*); Masson et al. (1980, *Aare Massif*); Steck (*Central Alps*); Escher (*Double fold*); Badoux (1972, *L'Argentine*); Nabholz (1975, *Flims*); Heim (1891, *Double fold*); Laubscher a. Bernoulli (1980, *Matterhorn*).

The TRANSMED Atlas; the Mediterranean Region From Crust to Mantle. Cavazza et al. (eds.) (2004). *Therein*: Stampfli a. Borel, The TRANSMED transects in space and time; constraints on the palaeotectonic evolution of the Mediterranean domain, pp. 53–80, Springer.

Pre-Mesozoic Geology of the Alps. v. Raumer a. Neubauer (ed.) (1993). Springer Verlag. 677 p. *Therein: Maps of the External Massifs.*

Der Geologische Aufbau Österreichs. Oberhauser (ed.) (1980). *Therein*: Janoschek a. Goetzinger (1969, *Molasse Zone*); Oberhauser (1980, *Erzberg*).

Reconciling Plate-Tectonic Reconstructions of Alpine Tethys with the Geological-Geophysical Record of Spreading and Subduction in the Alps. Handy et al. (2010). Earth Science Reviews, v. 102, pp. 121–158.

Geologie der Dolomiten. Bosellini A (1998). Athesia, Bolzano, 192 pp.

Geologie von Österreich 1. Die Zentralalpen. Tollmann A (1977). Deuticke, Vienna, 766 p.

Geologie der Alpen im Zeitraffer. Schuster R, Stüwe K (2010). Mitteilungen des Naturwissenschaftlichen Vereins für Steiermark, v. 140, pp. 5–21.

Selected scientific papers

Brückl E, Brückl J, Heuberger H (2001). Present structure and prefailure topography of the giant rockslide of Köfels. Zeitschrift für Gletscherkunde und Glazialgeologie, v. 37, pp. 49–79.

Cesare B, Mazzoli C, Sassi R, Spiess R, Sassi FP (2010). Beauty and complexity of metamorphism: case studies from the frontal part of the Adria microplate Rendiconti Lincei, v. 21, pp. 73–94.

Escher A, Masson H, Steck A (1993). Nappe geometry in the Western Swiss Alps. Journal of Structural Geology, v. 15, pp. 501–509.

Escher A, Masson H, Steck A (1988). Geologic cross sections of the western Swiss Alps. Memoires de Geologie Lausanne, v. 2, 11 pp.

Escher A, Hunziker JC, Marthaler M, Masson H, Sartori M, Steck A (1997). Geological framework and structural evolution of the western Swiss-Italian Alps. In: Pfiffner et al. (eds.) Deep Structure of the Swiss Alps. Birkhäuser, Basel, pp. 205–221.

Froitzheim N, Schmid S, Conti P (1994). Repeated change from crustal shortening to orogen-parallel extension in the Austroalpine units of Graubunden. Eclogae Geologicae Helvetiae, v. 78, pp. 559–612.

Grujic D, Mancktelow N (1996). Structure of the northern Maggia and Lebendum nappes, Central Alps, Switzerland. Eclogae Geologicae Helvetiae, v. 89, pp. 461–504.

Hergarten S, Wagner T, Stüwe K (2010). Age and prematurity of the Alps derived from Topography. Earth and Planetary Science Letters, v. 297, pp. 453–460.

Kastelic V, Vrabec M, Cunningham D, Gosar A (2004). Neo-Alpine structural evolution and present-day tectonic activity of the eastern Southern Alps: The case of the Ravne Fault, NW Slovenia. Journal of Structural Geology, v. 30, pp. 963–975.

Lardeaux JM, Schwartz S, Tricart P, Paul A, Guillot S, Bethoux N, Masson F (2006). A crustal-scale cross-section of the south-western Alps combining geophysical and geological imagery. Terra Nova, v. 18, pp. 412–422.

Leloup PH, Arnaud N, Sobel E R, Lacassin R (2005). Alpine thermal and structural evolution of the highest external crystalline massif; the Mont Blanc. Tectonics, v. 24, 26 pp.

Likorish WH, Ford M (1998). Sequential restoration of the external Alpine Digne thrust system, SE France, constrained by kinematic data and synorogenic sediments. In: Mascle et al. (eds.), Cenozoic foreland basdins of western Europe. Geological Society of London Spec. Publ., v. 134, pp. 189–211.

Linzer HG, Decker K, Peresson H, Dell'Mour R, Frisch W (2002). Balancing lateral orogenic float of the Eastern Alps. Tectonophysics, v. 354, pp. 211–237.

Missoni S, Gawlick HJ (2010). Evidence for Jurassic subduction from the Northern Calcareous 3 Alps (Berchtesgaden, Austroalpine, Germany) International Journal of Earth Science, DOI 10.1007/s00531-010-0552-z.

Mosar J, Stampfli GM, Girod F (1996). Western Préalpes Medianes Romandes; timing and structure; a review. Eclogae Geologicae Helvetiae, v. 89, pp. 389–425.

Nagel T, de Capitani C, Frey M, Froitzheim N, Stünitz H, Schmid SM (2002). Structural and metamorphic evolution during rapid exhumation in the Lepontine dome (southern Simano and Adula nappes, Central Alps, Switzerland). Eclogae Geologicae Helvetiae, v. 95, pp. 301–321.

Neubauer F, Lips A, Kouzmanov K, Lexa J, Ivascanu P (2005). Subduction, slab detachment and mineralization; the Neogene in the Apuseni Mountains and Carpathians. Ore Geology Reviews, v. 27, pp. 13–44.

Oberhänsli R (2004). Metamorphic structure of the Alps, Explanatory notes to the map with contribution of the authors in Ges., M.Ö.M., ed., v. 149, pp. 115–226.

Petrakakis K (1986). Metamorphism of high-grade gneisses from the Moldanubian zone, Austria, with particular reference to the garnets. Journal of Metamorphic Geology, v. 4, pp. 323–344.

Pleuger J, Roller S, Walter JM, Jansen E, Froitzheim N (2007). Structural evolution of the contact between two Penninic nappes (Zermatt-Saas zone and Combin zone, Western Alps) and implications for the exhumation mechanism and palaeogeography. International Journal of Earth Sciences, v. 96, pp. 229–252.

Rosenberg C (2004). Shear zones and magma ascent: A model based on a review of the Tertiary magmatism in the Alps. Tectonics, v. 23. doi:10.1029/2003TC001526.

Schmid SM (1975). The Glarus overthrust: Field evidence and mechanical model. Eclogae Geologicae Helvetiae, v. 68, pp. 247–280.

Schmid S (1983). Tektonische Profile der Helvetischen Decken im Reusstal-Querschnitt. In: Funk et al., Mechanismen der Gebirgsbildung. Eclogae Geologicae Helvetiae, v. 76, pp. 91–123.

Schmid SM, Pfiffner OA, Froitzheim N, Schönborn G, Kissling E (1996). Geophysical-geological transect and tectonic evolution of the Swiss-Italian Alps. Tectonics, v. 15, pp. 1036–1064.

Schmid SM, Kissling E (2000). The arc of the Western Alps in the light of geophysical data on deep crustal structure. Tectonics, v. 19, pp. 62–85.

Schuster R, Stüwe K (2008). Permian metamorphic event in the Alps. Geology, v. 36, pp. 603–606.

Steck A (1997). Geologic framework and structural evolution of the western Swiss-Italian Alps. In: Pfiffner et al. (eds.), Deep structure of the Swiss Alps. Birkhäuser, Basel pp. 205–221.

Tollmann A (1965). Geologie der Kalkvoralpen im Ötscherland als Beispiel alpiner Deckentektonik. Mitteilungen der Geologischen Gesellschaft, v. 58, pp. 103–207.

Zachos J, Pagani M, Sloan L, Thomas E, Billups K (2001). Trends, Rhythms and Aberrations in Global Climate 65 Ma to Present. Science, v. 292, pp. 686–693.

Sponsors

This book was sponsored by the Madrid based mining company SPmining (www.spmining.org). SPmining specialises in the investigation and promotion of natural resources from the African continent and operates with a visionary coupling of educational aspects in African countries, university research and traditional exploration. Strong focus is placed on state of the art remote sensing. As such, the bird's eye view of geology of the Alps as presented in this book serves as a symbol for the vision of SPmining and its activities. We thank for the close cooperation between Graz University and SPmining.

Page 285: The Haldensteiner Calanda (2805 m) is a mountain just west of the Rhine Valley near the city of Chur in the Glarus Alps. Geologically the mountain is part of the Helvetic nappes of Switzerland and is made up of Jurassic and Cretaceous limestones. The beautiful folds on the image at right are in its northwest face. Not far west of the mountain these rocks are overthrust by Permian volcanics along the Glarus Thrust.

Left: The Balmhorn (3698 m) in western Switzerland sits high above the Gastern Valley in the Bernese Oberland. Geologically it is part of the Helvetic nappe pile and it contains some of the most spectacular folds in the Alps (see p. 58).

Making of ...

Kurt's Impressions:
The idea for this project had been roaming around my head for years, but in absence of a good pilot, photographer and in fact a good plane, it simply never happened. Then, in 2008, I was with my family in one of the back corners of the Alaskan bush for geological field work, and saw some of Hombi's mountain photography in a wilderness lodge. It was there that my wife Karin reminded me that our old climbing friend Hombi had become a pilot in the last few years and might be an ideal partner for such a project. From that day onwards everything fell into place and every day that Hombi and I spent high above the Alps in his Piper Super Cub felt like a unique gift. What an unbelievable angle for a geologist and mountaineer like myself to see and understand our mountain range! Although every single day will be a lasting memory, there is one particular day that sticks in my memory: the 30th June 2009.

After a sticky hot night I woke up at dawn and crawled out of my sleeping bag in the dewy meadow next to the plane somewhere in the Upper Rhône Valley. Bitten by mosquitos, sweaty and barely awake, we climbed into the plane and started towards the thick clouds that covered the sky. 20 minutes later we circled only 100 metres from the summit of Matterhorn in brilliant sunshine and waved to the climbers through the open door of the plane at minus 5° C. An hour later we sorted photos during breakfast in a small cafe in the small township of Raron, before we refueled and started off for the 2nd time that morning. This time we were off to Mt. Blanc so that we could be in Gstaad for lunch ...

Hombi's Impressions:
In summer 2008, Kurt asked me if I was interested in taking photos for this project. I had known Kurt from the Himalayas and Alaska as a climber and very dynamic person for many years, but had not seen him for some time. In the meantime I myself had retired from my years as an active climber and had discovered my love for flying – it had been an old childhood dream. Taking photos of mountains from the air quickly had become a great passion.

The idea to cover the entire Alps, all the way from Nice to Vienna, and take good pictures immediately fascinated me. We planned four journeys of 3 to 4 days each, flying through all six countries that contain parts of the Alps, and split this up over

two summers. In total about 60 hours in the air were necessary, and we stopped at 20 different airfields, where we refueled, had to battle with customs formalities, wait for weather or rest for the night. Several times we camped right next to the plane.

The plane is a Piper Super Cub PA-18 built in 1957 and based in Bad Ragaz in the Rhine Valley. The mountain goat that is the symbol of the Swiss canton Grisons is painted on the cloth covered fuselage. The plane is perfect for mountain photography with fantastic slow flying properties. A 160 PS Lycoming Motor takes us safely to the highest peaks and through the deepest gorges. When taking photos, the door is wide open and it is icy cold in the cockpit. In the rear seat there are gale force winds and maps, cameras and navigation equipment can easily fly through the air.

For the photos we mostly used a Canon EOS 5 MarkII with a 24–105 mm tele-zoom lens. The camera endured the rough world of the cockpit very well. Taking photos at the same time as flying a plane takes a lot of concentration, and often steering is done with the knees to keep the plane at the right angle. When Kurt shouted, “Take the picture from 500 m higher”, it often took wild manoevers to bring the plane into the right angle. During each stop, the data was transferred to a laptop to make an initial check of the photos. At home many days of work on the computer followed. In particular the joining of the panorama images took a lot of time. In total we took about ten times as many photos as are shown in this book. Learning about the Alps from the air helped us to appreciate the unbelievable beauty of our mountain range.

Geological companion material to the book as well as the entire database of photos can be found on the interactive web page **http://www.alpengeologie.org**

Index

Index map of all photos

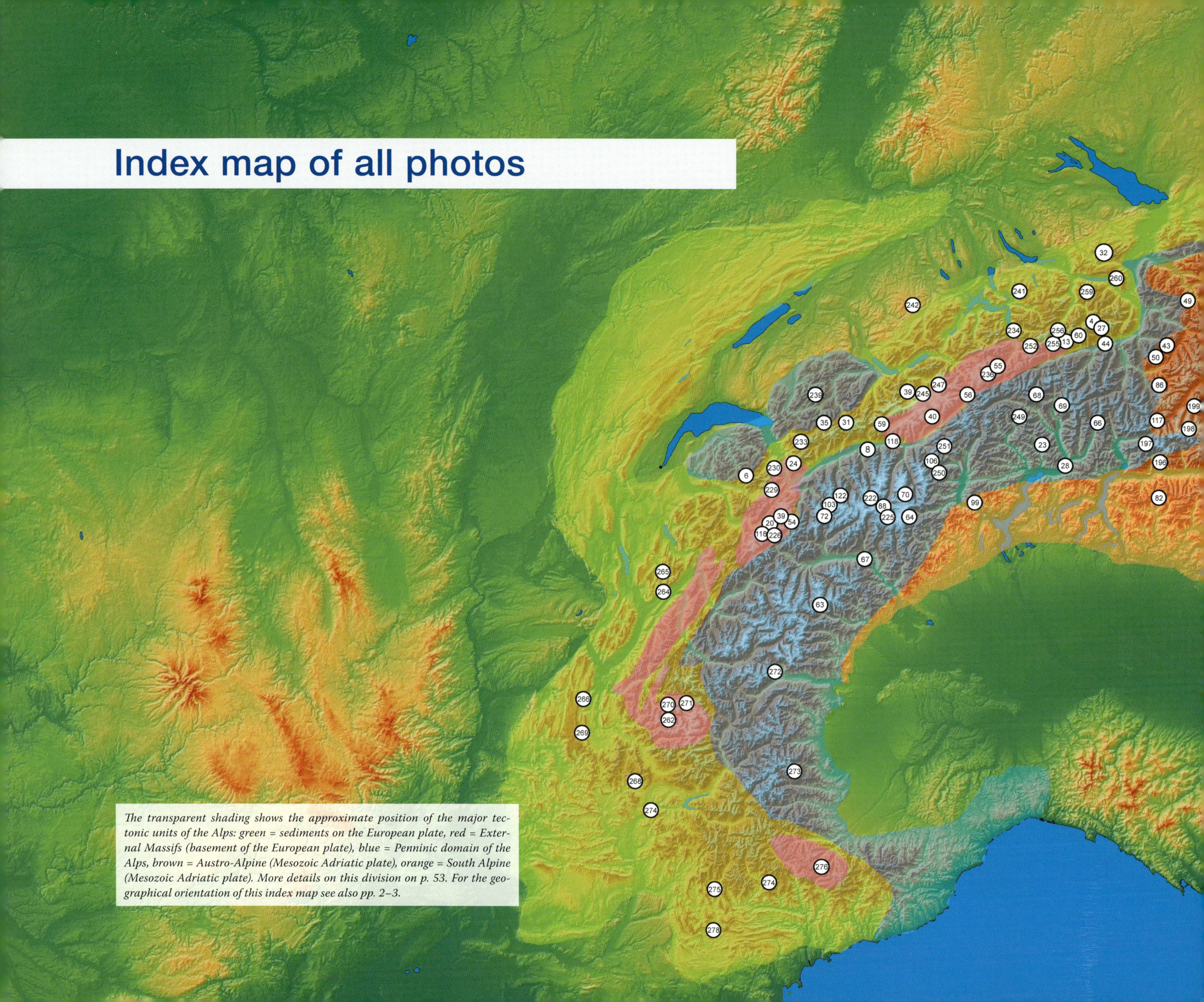

The transparent shading shows the approximate position of the major tectonic units of the Alps: green = sediments on the European plate, red = External Massifs (basement of the European plate), blue = Penninic domain of the Alps, brown = Austro-Alpine (Mesozoic Adriatic plate), orange = South Alpine (Mesozoic Adriatic plate). More details on this division on p. 53. For the geographical orientation of this index map see also pp. 2–3.

97
149
139
138
148
150
135
146
184
169
171
166
146
147
141
144
143
153
182
180
174
168
101
76
145
179
172
173
185
177
114
78
9
161
160
152
176
75
122
191
188
163
133
158
159
38
126
164
130
112
156
105
95
192
187
98
131
104
128
216
217
94
15
154
108
209
95
136
200
120
219
201
202
218
207
204
36
208
221
220
123
81
111
90
93
210
211
212

The cover picture shows the Alpstein massif in the Appenzell Alps of eastern Switzerland. More details about this photo are available on pp. 32–33. The eight small pictures on the back cover show from top left to bottom right: the Tödi, highest mountain in the Glarus Alps (p. 254); geological sketch of Tödi (p. 254); Mont Aiguille in the Vercors Massif of France (pp. 266–269); the Pala Group in the Dolomites (pp. 216–221); geological map of the Mont Blanc massif; the summit of Mont Blanc (p. 226); fold in the Dent de Morcles (p. 24) and the Styrian Erzberg (p. 144).

ISBN 978-3-7059-0343-2
1st English edition 2012

ISBN of the 3rd German edition: ISBN 978-3-7059-0314-2

Tel.: 03151-8487, Fax: 03151-84874
e-mail: verlag@weishaupt.at
e-bookshop: www.weishaupt.at

Printing and binding: Theiss Printing Press GmbH, A-9431 St. Stefan.
Printed in Austria.